# THE *COSMOGRAPHIA* OF BERNARDUS SILVESTRIS

Records of Western Civilization

# RECORDS OF WESTERN CIVILIZATION

A series of Columbia University Press

# THE *COSMOGRAPHIA* OF BERNARDUS SILVESTRIS /

*A Translation with Introduction and Notes by Winthrop Wetherbee*

*COLUMBIA UNIVERSITY PRESS*
*NEW YORK*

Columbia University Press
New York   Oxford

Publication of this book has been made possible by the Hull
Memorial Publication Fund of Cornell University.

Library of Congress Cataloging in Publication Data
Bernard Silvestris, fl. 1136.
    The Cosmographia of Bernardus Silvestris.
    (Records of civilization: sources and studies,
    no. 89)
    I. Wetherbee, Winthrop, 1938–     ed.
II. Title.   III. Series.
PA8275.B25C613.     128     73-479
ISBN 0-231-03673-6
ISBN 0-231-09625-9 (pbk.)

*For C.H.W.*

Records of Western Civilization is a new series published under the auspices of the Interdepartmental Committee on Medieval and Renaissance Studies of the Columbia University Graduate School. The Western Records are, in fact, a new incarnation of a venerable series, the Columbia Records of Civilization, which, for more than half a century, published sources and studies concerning great literary and historical landmarks. Many of the volumes of that series retain value, especially for their translations into Enlgish of primary sources, and the Medieval and Renaissance Studies Committee is pleased to cooperate with Columbia University Press in reissuing a selection of those works in paperback editions, especially suited for classroom use, and in limited clothbound editions.

Committee for the Records of Western Civilization

Malcolm Bean, ex officio

Caroline Walker Bynum

Joan M. Ferrante

Robert Hanning

Robert Somerville, editor

# *Preface*

Bernardus Silvestris has received far less attention than either the merit or the historical importance of his writings would justify, and it is with a real sense of responsibility that I offer this version of his major work. Consultation with learned friends has suggested to me that little in this volume is likely to be accepted as definitive, and I am particularly aware of how far short my rendering of the *Cosmographia* falls of the beauty and allusiveness of the original. But I think that it is worth while to make available even in this imperfect form the work of a writer whose influence gained for him the status of *auctor,* and who has been found worthy of comparison with Dante, Bruno, and Shelley.

Though I have attempted to characterize the theological ideas which pervade the *Cosmographia,* and to document with some thoroughness the classical and medieval components of Bernardus' philosophical thought, I have sought above all to present his cosmogony as a work of the imagination, and to point out the essentially poetic character of the patterns of allusion and analogy which are the source of its real coherence. In this I have followed Professor Theodore Silverstein who showed, in an article which remains a valuable starting point for any consideration of Bernardus' work, that in an age when theological controversy was intense and rationalism was regarded with mistrust, "what philosophy could not do, poetry might."

The project originated as the suggestion of Professor Paul Piehler, who has provided valuable advice and encouragement along the way, and a fellowship from the American Council of Learned Societies enabled me to complete it. Miss Meridel Holland gave an earlier version of the translation a careful reading and suggested some important revisions. I am grateful also to Professor George Economou for his interest and encouragement; to Professor Brian Stock and Mr. Peter Dronke for making available to me unpublished materials in their possession; to the Instituto Británico of Seville, and to Professor Juan Gil and the staff of the Seminario de Latin, Universidad de Sevilla, for their generous hospitality during the time when I was completing the book. Professor W.T.H. Jackson, General Editor of the Records of Civilization, has been patient with a project which has moved to completion through slow stages, and the translation in its final form incorporates a number of his suggestions.

Brian Stock's *Myth and Science in the Twelfth Century: A Study of Bernard Silvester* (Princeton, 1972) appeared when my work was in the final stages of preparation. Thus I have been able to note only the most important points of contact between Professor Stock's treatment of Bernardus and my own, but am happy to be able to recommend this fine study, rich in new insights and new material, as a contribution of the first importance toward that just appreciation which Bernardus has been so long denied.

*Winthrop Wetherbee*

*Ithaca, New York*
*December, 1972*

# Contents

# Introduction

## 1.
## The Cosmographia and the Twelfth-Century Renaissance

The main features of the intellectual movement of the late eleventh and early twelfth centuries are well known. Among them are the revival of dialectic and the theological speculations of Anselm; a new coming to terms with the themes and mythological apparatus of classical poetry, represented by the work of Hildebert of Le Mans, Baudri of Bourgueil, and others; and the activity of the pioneering transmittors of Arab science, whose work stimulated a widespread interest in the study of the natural world. The development in urban centers of cathedral schools, largely devoted to practical training in law, rhetoric, and logic, gave an increasingly worldly orientation to academic pursuits, and the absence of the constraints of traditional monasticism encouraged an expansion and a liberalizing of school curricula.

From one point of view these developments reflect a growing interest in the situation of man in the world, and in the development of a secular culture. The emergence of the "intellectual" as a type corresponds to a new concern with the dignity of work of many kinds, a rise in the status of technology relative to the traditional liberal arts and an awareness of man's power to impose a discipline of his own, mechanical and cultural, on that *fabrica mundi* of which he is an integral part.[1] But the advances in intellectual freedom and self-consciousness in the period do not represent simply an abandoment of the essentially spiritual concerns of earlier medieval thought. As Olaf Pederson observes, with reference to the impact of the thought of Anselm, once religious thought was conceived as the activity of a "fides quaerens intellectum," the way lay open to an employment of all the resources and possibilities of human knowledge in this quest.[2] And not the least of the discoveries of the twelfth century is the science of theology.[3]

As several recent surveys of the period have pointed out,[4] the thought of Anselm heralds a new concern with the relations between God and man, the presence of the divine in nature—with the earthly experience of the God-man, Christ, as a focal point for meditation, and with the more general implications of the cosmic and psychological renewal effected by the Incarnation. And a unifying factor among the many forms of religious expression in the period is a tendency to view this relationship in structural terms. Treatises *de anima* proliferate;[5] new lyric forms are devised to articulate new refinements of devotional feeling (and find, of course, a complement in the refinement of erotic emotion in the emerging secular love-lyric);[6] the principles of cathedral architecture are made to express mystical ideals.[7] The study of classical literature, as well, from the most rigorously scientific analyses of the platonist cosmology (which remained the authoritative "model" of universal order) to the increasing tendency to allegorize the gods of the ancient pantheon as "Mittelsmächte" in the cosmic process, serves finally to demonstrate how, in the phrase of Wolfram Von den Steinen, "der Eintritt des Göttlichen die Natur verwandelt."[8]

It is within this basically theological context that the *Cosmographia* of Bernardus Silvestris must be considered. The *Cosmographia* is a landmark of twelfth-century humanism.[9] Its theme illustrates perfectly the ambitions and resources of the Platonism of its day, and its literary form and character reflect one of the earliest significant confrontations between a modern European author and the classical tradition. Implicit in its allegory is a subtle critique of the sacramental, as well as the psychological and moral implications of the Platonist cosmology, and of the new sense of the autonomy and value of universal life which the twelfth century saw reflected in this cosmology; at the same time it is in a special sense an "epic," a definitive and heroic characterization of human experience.

Philosophically, the *Cosmographia* expresses that concern to expand the limits of rational speculation, to affirm both the dignity of man and the dignity of that natural order with which man's nature, even in its fallen state, exhibits profound affinities, which is the essence of medieval humanism. Its cosmology and the pattern of its analogy between the larger universe or "megacosmos" and the human "microcosmos" are owed to Plato's *Timaeus*,[10] and it provides as well a brilliant distillation of the tradition of neo-Platonist encyclopedism, as represented by Calcidius, Macrobius, Martianus Capella, and Boethius, with which the *Timaeus* was inextricably involved in the Middle Ages. Its psychology and its account of human destiny also reflect a careful exploitation of certain features of Christian neo-Platonism, as developed by Augustine, by Bernardus' twelfth-century predecessor Hugh of St. Victor, and most strikingly in the system of Johannes Scotus Eriugena. And its immediate inspiration was the activity of the twelfth-century scientist-humanists: the spread of the study of medicine, the translation and promulgation of Arabic scientific texts, and the development of a coherent rational Platonism by that group of teachers and philosophers conventionally associated with the School of Chartres.

But the *Cosmographia* is also literature, the work of a poet as well as a philosopher. It appeared toward the middle of the twelfth century, at a time when the intellectual progress of the earlier years of the century was giving rise to new attitudes toward the study and creation of literature in the schools, and gradually influencing vernacular culture as well. Its author, in an age when scientific and literary studies were often closely intertwined, is a particularly striking example of their combination. He was the friend and disciple of Thierry of Chartres, and perhaps the author of commentaries on Plato and Martianus Capella, as well as the translator of an Arab astrological treatise. At the same time he was a poet and professional rhetorician, the teacher of such *litterati* as Matthew of Vendome, and his *Cosmographia* provided a model for the great allegorical poems of Alain de Lille. His *Mathematicus*, too, enjoyed a wide reputation, and his contribution to the scope and the codification of mythography, if we accept the attribution of the commentary on Martianus, and of another on the first six books of the *Aeneid*, was of the utmost importance for later Latin and vernacular poetry. And it is finally as a poem that the *Cosmographia* must be understood.

Bernardus' centrality can be seen in the fact that scholars of widely divergent interests have found him a particularly representative figure. Theodore Silverstein, in an article which was the first serious study of the intellectual background of the *Cosmographia*,[11] has demonstrated how the work crystallizes in poetic language and imagery complex philosophical and theological concepts which might have aroused serious criticism (and in fact did so in the case of Bernardus' fellow Platonist Guillaume de Conches) if presented in prose argumentation. Tullio Gregory has made it possible to see in the *Natura* of the *Cosmographia* the distillation of a generation of rich speculation, and the focusing of its implications on the moral and psychological situation of man.[12] Others have shed light on major developments in physical theory and medicine which are illustrated in image in Bernardus' characterizations of Physis and Urania.[13] From a different vantage point literary scholars have recently presented suggestive theses relating Bernardus' conception of the silva, the physical (and by implication the psychological) chaos from which all created life is formed and articulated, to the "gaste forest," the symbolic wilderness of Arthurian romance, and have stressed Bernardus' formative influence on the twelfth-century conception of epic poetry.[14] And Alfred Adler has detected the embryonic expression of the demand of an increasingly self-conscious courtly and urban society for refinement and dignification in Nature's great appeal to God "that the universe be more beautifully wrought." [15]

But while the *Cosmographia* thus serves to crystallize in poetic form the main themes of twelfth-century humanism, it also reveals its author's sensitivity to other aspects of the thought of the time; some of these darken and complicate its presentation of the human condition, while others serve finally to create a deeper spiritual unity. It is an oversimplification to see in the *Cosmographia*, as Helen Waddell and Edmond Faral have done, simply an exuberant expression of joy and confidence in man and his nature, a major step in the direction of later and bolder "naturalisms." [16] In the astronomical portions of the *Cosmographia* and the darker hints of the *Mathematicus* we must recognize the recurring pessimism which was inseparable from the widespread interest of the twelfth century in astrology and divination.[17] Cosmology and astrology were closely allied in this period and in Bernardus' writings; as the former tended increasingly to emphasize the autonomy of the cosmic order, the latter translated the operation of this order into terms of its determining influence on the course of human life. Under the influence of these concerns Bernardus raises at several points in the *Cosmographia* the question of the role of fate in human affairs. On the other hand, the introduction of Christian neo-Platonism necessarily entailed, in the twelfth century, contact with an emanationist view of life for which the facts of formal order and continuity in nature are at best accidental phases in, and at worst obstacles to the participation of human life in the great movement of procession and return in which the true relationship of creation with God consists. And Bernardus' late-antique sources also exposed him to a non-Christian neo-Platonism, for which life is an endless struggle between spirit and flesh, one seeking always to return to its divine source, the other tainted by the *malignitas* of materiality and posing the threat of dissolution.

It is largely in the tension between what we may call the "conventional" Platonism grounded in the *Timaeus* and the potentially contradictory implications of these heterogeneous elements in Bernardus' thought that the meaning of the *Cosmographia* consists. It is a meaning which does not yield itself easily, for Bernardus' ultimately affirmative vision of man's spiritual destiny rests in almost unresolved coexistence with a genuine pessimism about the obstacles to spiritual advancement presented by the nature and condition of man. One purpose of this introduction will be to demonstrate the fundamentally poetic character of the synthesis achieved in the *Cosmographia* and the oblique manner in which Bernardus conveys his deepest insights. It is necessary first, however, to review briefly the traditions in philosophical and literary studies on which Bernardus' major conceptions depend, and the various sources which contributed to the synthesis of the *Cosmographia*.

## 2.

## *The Development of the Idea of Nature in the Twelfth Century*

The *Cosmographia* opens before the beginning of time, with Nature's appeal to Noys, divine providence, on behalf of Silva, the material of created life, which yearns to come into existence. In response to her plea, Noys creates the greater universe, fills the earth with animal and vegetable life, and then charges Nature herself with the task of convoking Urania, celestial *ratio,* and Physis, representing the principles of physical life, to fashion the soul and body of man, which Nature is then to join together. This the goddess does, and the work ends with an account of the fashioning of the human body, the lesser universe.

Nature, the protagonist of the cosmic drama, is in many respects a discovery of the twelfth century. Those scholars with whom we associate the intellectual renaissance of the period seemed to themselves to be asserting new and important truths in claiming that the universe was an object of study worth considering for its own sake. And it is certainly true that the history of cosmological thought in the earlier Middle Ages, apart from such practical concerns as the establishing of the church calendar, is extremely uneven. Scientific and philosophical consideration of the universe and its operations were in general contained and subsumed, as indeed they were largely to be in the twelfth century itself, by the conviction that the importance of *naturalia* was necessarily supernatural, transcending any coherence or correspondence they might exhibit among themselves. Allegorization "crowns" the numerous borrowings from ancient cosmological writings in the *De naturis rerum* of Isidore of Seville, which combines encyclopedism with a meditation on the plan of salvation.[18] As Gregory observes, there is little in the intellectual character of such treatises as the *De universo* of Hrabanus Maurus to suggest the existence of autonomous laws and self-sufficient forces in created life; the universe which he presents is a compendium of allegorical and mystical significances.[19] Even the great scheme of Eriugena, with its hierarchy of operative *naturae* and its insistence on the integrity and goodness of creation at all levels, is finally only a systematizing of this radically mystical view of natural existence, an articulation of the completeness with which divine reality pervades and gives meaning to all life. This is not, of course, to say that natural philosophy and the *quadrivium* were extinct in the early Middle Ages; Eriugena was himself an astronomer and physicist as well as a theologian and speculative philosopher,[20] and in his commentary on Martianus Capella we can see the survival of that neo-Platonist encyclopedism which was to play a major role in the intellectual revival of the eleventh and twelfth centuries. The many commentaries of Remigius of Auxerre, and a series of later glosses on the cosmological hymn, the "O qui perpetua," of Boethius' *De consolatione philosophiae* kept alive in some form the late classical tradition of critical interest in the Platonic cosmology.[21] But from the time of Eriugena forward there was a definite tendency to regard such studies with suspicion, and in proportion, it would seem, as interest in the universe and the sciences gained ground in the schools the reaction against them became more articulate. In the *Opusculum contra Wolfelmum* of Manegold of Lautenbach, written about 1080, we may see the essence of the anti-scientific polemic, an attack on the preoccupation of philosophers with "seeking out the natures of things" building to a scornful dismissal of such thinkers as incapable of imagining a substantial existence beyond this world.[22] Peter Damian stresses God's omnipotence and his power to suspend whenever he wishes the force of necessity in cosmic existence, thus confounding the assumptions of the philosophers and dialecticians, and he even goes so far as to take issue with St. Jerome over the question of God's power to restore to a woman who has lost it the purity of virginity.[23] Such attitudes survive in the twelfth century, in the attacks of Guillaume de St. Thierry, Gautier de St. Victor, and others on the seeming challenge presented by the *physici* to the authority of Scripture.

But in the early twelfth century a number of developments coincided which, for some at least, put the study of the sciences in a new light. Medical study had been revived in Italy, and in particular the School of Salerno had gained a wide reputation by making available Greek and Arab learning in this field.[24] From this initial contact, and through the growing accessibility of Arab scholarship to Latin scholars in Spain, there grew an intense interest in other aspects of the learning of these alien cultures. Astronomy and physics, separate at first but tending steadily toward coalescence in a single "philosophia mundi," received a new stimulus.[25] With the influence of these encounters was combined that of the technological demands of a European society in the process of achieving a new stability, and hence interested from a practical point of view in the contributions which a fuller understanding of the "principia rerum" might make to metallurgy, architecture, agriculture, and other fields.[26]

However, the spread of the new science was slow, and the signs of its effect in northern Europe in the early twelfth century are new attitudes rather than new discoveries: an insistence upon the value of rational investigation, a tendency to elaborate schemata classifying the sciences in terms of their contribution to a single coherent wisdom, and to glorify the vocation of the scholar as leading to a unique and indispensable conception of reality. Working with largely the same *auctores* accessible to Remigius, Adalbold of Utrecht and Bovo of Corvey, thinkers like Guillaume de Conches place a new importance on understanding the workings of nature.[27] A striking instance of the new attitude is Guillaume's scorn of those thinkers who evade their responsibility to seek knowledge by referring to God's miraculous power anything which they do not understand. For Manegold of Lautenbach the history of miracles, suspending the workings of the cosmic order again and again, had been a decisive argument against the presumption of the "philosophi" to explain the working of the divine: "so often has the wonted course of nature been thwarted that now nature herself can scarcely trust her own powers," he notes with satisfaction.[28] But Guillaume, convinced of the necessity of finding the rational laws and causes of things, rejects the literal truth of the biblical account of the birth of Eve, and having relegated this to the level of allegory, defends his right to explain how that happened which faith commands us to believe.[29] The same trust in the doctrines of science enables him to reconcile with Christian belief Plato's myth of the pre-existence of human souls, each in its star. This, says Guillaume, must be understood "causaliter, non localiter," as a metaphor for the influence of the heavens on earthly life.[30]

This last illustration, even as it reveals the beginnings of a "concordisme doctrinale" between Christian and Platonic doctrines,[31] suggests another important element in the new scientific attitude, the remarkable spread of astrology and divination. The idea that the events of earthly life were governed and predetermined by the orderly disposition and activity of the heavenly bodies and could, in part, be foreknown through the careful analysis of celestial phenomena was from the beginning inseparable from propaganda for the new science, and significantly affected the celebrated humanism of the twelfth century.[32] The notion, articulated in many ways in the period, that self-knowledge depends on a knowledge of the universe, is inseparable from the conviction that a man may in part determine the course of his own life by exploiting the knowledge implanted in the stars, and learning what to do and what to avoid. And the many references to the Ciceronian ideal—dramatized in the elaborate allegory of Martianus—of a union of *sapientia* and *eloquentia,* or, in practical twelfth-century terms, of *quadrivium* and *trivium,*[33] was accompanied by a recognition of the deficiency of the Latin *quadrivium* in comparison with that of the Arabs, who had, it was felt, rightly placed a special emphasis on the sciences and their role in helping man to sense the larger pattern of life.[34] Adelhard of Bath, in the *De eodem et diverso,* extols the power of the Arts to guide the soul in its earthly journey; they teach her to recognize her special relation to the rest of creation, to know the nature and intuit the divine pattern of the universe. For the soul's basic affinity is with the divine *rationes* of things, and in her pure precorporeal state she "examines not only things in themselves but their causes as well, and the principles of their causes, and from things present has knowledge of the distant future."[35] For the most part Adelhard's language echoes Boethius, Vergil, and the *Asclepius,* but the thoughts expressed are largely those of Firmicus Maternus and the astrologers.[36] Elsewhere in the work, when Philosophy commends the Liberal Arts to the author, she declares of Astronomy that "if a man can possess her he will cease to doubt, not only regarding things close at hand in the earthly sphere, but even things past and to come."[37]

Astrology plays a complex role in twelfth-century attitudes toward the situation of man. It implies a determinist view of life which inevitably generated pessimism and what must seem to us an absurd preoccupation with the purely divinatory aspects of astrological calculation; but viewed in another way its very status as a science implied the possibility of transcending the stars by mastering the knowledge they offered, and, if not evading one's fate, at least accepting it as one's human lot and meeting it nobly, thereby asserting a certain superiority. It could be construed as yet another of the many ways in which the creation exists to serve man, and the idea of comprehending it could suggest a new sense in which man's mind mirrors the totality of nature, reflecting in this its affinity with the all-disposing wisdom of God. We will see striking reflections of both the optimistic and the pessimistic sides of the astrological mentality in the poetry of Bernardus.

But astrology and its practical implications were only one reason for the study of the universe, and only gradually became separated from the traditional Platonist idea that the universe and its harmonious order reveal the ideal patterns on which human life and virtue are founded, and are thus a means of access to the mind of God and the essence of human psychology. This view had been given vivid expression in Boethius' magnificent philosopher's prayer, the "O qui perpetua," a brilliant distillation of the cosmology of Plato's *Timaeus*. The *Timaeus* itself was the single most authoritative model of the cosmic order, and from the beginning the assimilation of new scientific ideas was marked by a renewed and increasingly sophisticated interest in Plato's cosmology. At a time when the sciences were seeking autonomy, Plato's characterization of the universe as a single, fully realized, and animate being was of tremendous importance, for it seemed to sanction the study of causality and analogy within the cosmic framework as manifestations of the "aurea catena" or hierarchy of cosmic powers which descends from the Platonic Demiurge and expresses its will.[38] Once granted an autonomous and coherent existence, the secondary causes of universal life could be studied as objects in themselves, expressions of a providential order, harmonized and sustained by the power of the world soul.

This world soul was the subject of crucially important discussions; it was regarded now as the Holy Spirit itself, the direct expression of God's benevolence, now as a principle of nourishment, an "igneus vigor," source of the unflagging vitality and intelligence of the heavenly bodies, and of the perpetual generation of earthly life. The former view was put forth over-boldly by Guillaume de Conches, who was taken to task by Guillaume de St. Thierry and subsequently modified his position.[39] The latter was the result of a fruitful union of Plato's cosmology with Stoic physics, and in particular with the concept of the "ignis artifex" as found in Cicero's *De natura deorum* and the *Quaestiones naturales* of Seneca, augmented in certain details by the *Asclepius*.[40] Further reinforcement for this conception was provided by the growing influence of the *Introductorium maius* of Abu Ma'shar, in which the influence of the firmament and the planets on physical and psychological existence was explained in detail.[41] Gregory has traced the process whereby, once the association of the world soul with the generation of life had been tentatively established, there gradually emerged the idea of a strictly cosmological principle, autonomous within its sphere, and responsible for all creation within the universe.[42] Thierry of Chartres, glossing the opening verses of *Genesis* "secundum physicam," calls fire "quasi artifex et efficiens causa," in the generation of elemental life, thereby restricting the activity of God to the creation of the elements themselves.[43] This creative fire, descending "ex quadam vi" to ensure the renewal of life in the sensible world, is identified by Hugh of St. Victor with *natura*.[44] Hermann of Carinthia, translator of the *Introductorium*, inspired both by Thierry and by first-hand familiarity with Arab doctrines, calls this same "nature," which he conceives as the special vehicle whereby divine life imparts itself to creation, "the power, proper to all created life, of reproducing and sustaining itself." [45] This aspect of nature, as preserving a "legitimum societatis foedus" between the heavens and earthly life, is developed by Dominicus Gundisalvus, largely on the basis of Hermann's work.[46] Finally, in the anonymous hermetic *De septem septenis*, Nature emerges full-blown: [47]

A created spirit, that is, a universal and natural motion, joins form, invisible in itself, with matter, invisible in itself, so that what is composed of these emerges as an actual visible substance. This spirit, this universal and natural motion, embracing the four elements as a sort of primal matter, the initial fusion of material existence, is diffused in the firmament through the stars, and in the sublunar world through fire, air, water, and earth. . . .

If we emphasize the synthesizing and containing function of the power here defined, we have essentially the Nature who plays so important a role in the *Cosmographia,* and her discovery has been claimed as the great event of the "renaissance" of the twelfth century. For the discovery of Nature, a power identified precisely with the preservation of life and order, and the obedience of all creation to cosmic law, meant the discovery of a new, more profound meaning in form and order themselves; her autonomy, like the vitality of the world soul, was imparted to all aspects of life in her domain. The perfection and coherence of the universe, as well as its relation to the divine source of life, were to be valued, and could, moreover, be understood as expressive of qualities to which man could discover correspondences in his own nature.[48] Man, like the universe, lives and moves through the interplay of rational and irrational forces; his affinity with nature imposes upon him the responsibility of self-government, as the rational firmament governs the wandering stars, and the measure of his integrity is the extent to which he can achieve in his own life the stability and regularity of the universe at large. Aesthetically too the discovery of nature had a considerable effect. As Chenu suggests,[49] the appreciation of nature helped to evoke a new appreciation of realism in art, a new interest in aesthetic form, a new preoccupation with the archetypal implications of myth and the themes of classical literature. Lyric poetry of a newly serious and analytical kind appears, testifying to a new recognition of the intrinsic dignity of human emotion. For nature is the source and object of human desires, and they, too, conform to her laws, albeit imperfectly. Ethics assume increasingly a natural, rather than a theological basis;[50] for man, like the universe, is composed of elements, his nature expresses itself through temperaments and humors which have a certain necessity and inevitability. *Amor, ira, concupiscentia* are capable of evil, but they are in themselves natural, and therefore to be valued in relation to intention and human frailty, rather than simply on the basis of their good or ill effects. Human experience, in short, becomes increasingly important, and comes, moreover, to be viewed from a new perspective: not only in terms of moral and spiritual awareness and enlightenment, but as reflecting the relations of a stabilizing and vivifying nature with a necessity, a psychological and physical intractability conceived as a basic condition of life.

*3.*

*Nature and Allegory*

"Philosophy," says Thierry of Chartres in the preface to his *Hepta-teuchon*, "is the love of wisdom, and wisdom is the coherent understanding of the true nature of existence." [51] With this conception of wisdom in mind, Thierry had compiled his "Seven Books of the Arts," a compendium of the authoritative texts in each of the Liberal Arts, probably to serve as the foundation of the curriculum of the School of Chartres. The idea that the "cohaerentia artium" provides an authoritative model of reality was not, of course, a new one; it had provided the rationale for Martianus Capella's allegorical "Marriage of Philology and Mercury." It was implicit in Macrobius' characterization of Vergil as versed in all branches of learning, so that his verse, its imagery and language informed by his knowledge, becomes a counterpart of the "divinum opus mundi" itself.[52] Firmicus Maternus had declared that a true mastery of the Arts was a confirmation of the divine origin of human understanding,[53] and Boethius had extolled the power of the *quadrivium* to illumine "the darkened eye of the mind" and direct its gaze to higher things.[54] But in the twelfth century, and in conjunction with the discovery and growing appreciation of Nature, it achieved a new importance.

For Thierry and his fellow "Chartrians," the most perfect illustration of the ideal fusion of learning and its expression, *sapientia* and *eloquentia,* was the body of texts in which the great *auctores* had embodied their wisdom. To Macrobius, Martianus Capella, and Boethius were devoted commentaries in which the resources of all the Arts, from grammar to the most complex aspects of music and astronomy, were used to expose neo-Platonist profundities and to define their relation to the doctrines of Christianity. A framework for such investigations was provided by the *Timaeus* of Plato, the philosopher whom Macrobius had called "the repository of truth itself." I have already suggested the importance of the *Timaeus* in promoting the study of natural law and the structure of the universe. In the broader plan of study defined by the *Heptateuchon* the *Timaeus* provided a means of coordinating the insights of other authors, translating their imagery into cosmological and psychological terms and systematizing their mythology. Plato's cosmos was understood as providing the essential context of the writings of all the great *auctores,* and in relation to this model the study of literature in general assumed a new importance. It was a commonplace in the twelfth-century schools that the *auctores* had used *involucra* of imagery to veil their most profound utterances from the eyes of the ignorant and profane. In this they had followed the practice of Nature herself, who, as Macrobius had declared in a famous passage, "just as she has withheld an understanding of herself from the uncouth senses of men by enveloping herself in variegated garments, has also desired to have her secrets handled by more prudent individuals through fabulous narratives." [55]

The concept of the *involucrum* as an element consistently present in the works of the *auctores* had been known to the early Middle Ages; implicit in the commentaries of Remigius and Eriugena on Martianus, it is clearly enunciated in a well-known poem of Theodulph of Orleans.[56] Sigebert of Gembloux finds occasion in his *De scriptoribus ecclesiasticis* for generous praise of the mythographer Fulgentius, "who has interpreted the whole system of *fabulae* according to the philosophy they express, with reference either to the order of things or the moral conduct of human life." [57] And Bovo of Corvey, inspired by Boethius' "O qui perpetua," anticipates Chartrian treatments of the Timaean cosmos, declaring that God, like an artist with the whole plan of his work in mind, has willed the perfection of the cosmos "in toto et in partibus suis" as a revelation of his wisdom.[58] But in the twelfth century the idea is developed systematically and made to bear a new weight of meaning. Abelard, in his theological writings, speaks freely of *involucra* as a mode of allusion common to the prophets and the philosophers, employed by Christ in his parables and capable even in pagan writings of intimating things so mysterious as the persons of the Trinity, and he cites Plato as especially favored with the power thus to convey intuitions of truth.[59] Guillaume de Conches, while willing to assign a theological validity to Plato's images, is more precise in his handling of the concept of *involucra,* or, as he more commonly calls them, *integumenta,*[60] less interested in their possible affinities with the mode of prophecy and more in defining the range of materials which lend themselves to interpretation. His commentaries on Plato and Boethius contain numerous discussions of the intricacies of grammar, and conventional mytho-

graphical moralizings, etymological and allegorical, reflecting his conviction that "it is not to be believed of such philosophers that they
would have inserted anything superfluous, or serving no purpose, into
such perfect works." [61] There are also highly original glosses on particular mythological allusions, incorporating psychological and literary
references which enrich and give continuity to Guillaume's over-all
interpretation. Finally there are sustained expositions of the cosmological and theological conceptions of the *auctores,* justifying where
possible and rejecting when necessary positions not obviously consonant with Christian doctrine.

In all of this, and above all in the conviction that every detail of
the writings of the great *auctores* bears reference to an underlying
"sensus mysticus," there appears a new sense of the significance of scientific knowledge, coupled with the expressive capabilities of imaginative literature; the continuity and authority of Plato's cosmology
are understood as reflected in the forms and themes of literature.
Such a critical approach is far removed from the assumptions of the
traditional "accessus ad auctores," and approximates in many respects
the Augustinian conception of the integrity and transcendent reference of the language of the Scriptures.[62] The comparison can only be
developed within definite limits, for despite the general revival of
philosophical and scientific studies, few were willing to go so far as
Guillaume and the Chartrians in their reliance on the wisdom of the
*auctores.* But their practice does reflect certain more general tendencies of the early twelfth century. For the Platonism on which
Chartrian criticism was based was only one of several Platonisms
abroad in the period. There is no doubt that a certain osmosis took
place between the Chartrian view of universal reality, centered primarily on the implications of structure and analogy, and the neo-Platonism which emerged, largely under the influence of the pseudo-
Dionysius and Johannes Scotus Eriugena, in such authors as Honorius
Augustodunensis and Hugh of St. Victor, and which conceived the
universe as charged in all its parts with mystical significances which
coalesce in a "theophany," a manifestation of God.[63] At St. Victor,
and among certain Cistercian thinkers, there appeared a striking and
consistent concern, as Robert Javelet has demonstrated, to synthesize
the newly circulating cosmological and physical notions and sub-

sume them to an essentially mystical view of the universe as "un
échelon harmonisé à toute anagoge vers l'Eternel." [64] Such a view had
of course the effect of deemphasizing the importance of continuity
and regularity within the cosmos itself, but at the same time it ex-
ploited these, and could greatly enrich the suggestiveness of the many
analogies between natural and divine reality; in its fully developed
form it was itself, as Chenu has indicated, a kind of science,[65] and its
insights tended to find expression, like those of the more strictly "scientific" Platonists, in structural terms. An illustration of this is the development of such concepts as that of the "sacrament" in the thought
of Hugh of St. Victor. Hugh, though convinced of the necessity of a
sound training in the Liberal Arts for the would-be exegete or theologian, is consistent in opposing the preoccupation of so many of his
contemporaries with mere "literature," and with the intrinsic implications of cosmology. Above all he repudiated the attempt to find religious truth in the *auctores*.[66] Though willing to acknowledge traces of
the divine handiwork in the Platonic cosmos, he tends to regard the
"sapientes huius mundi" as trapped in the closed system of their own
ideas, committed to searching in nature for a reality which is beyond
nature.[67] At the same time Hugh is acutely sensitive to the possible
implications of such Platonic figures as the world soul, which he reduces to a metaphor for the potential comprehension of all reality by
the human mind,[68] and his analysis of the work of the six days includes a number of reflections on the "sacraments" implicit in the process of creation. As matter was first created "in forma confusionis," so
the soul, while in a state of sinfulness, "in tenebris quibusdam est et
confusione." [69] But as the light of truth was present before the ordering of the heavens, so it is present before the moral ordering of the
soul; with its help the soul gains an awareness of good and evil, and
the "sun of justice" begins to shine. Man is "a great mass of desires,"
like the primordial chaos, but his reason is informed by a "motus im-
mortalis vitae," and is thus able to mediate between the celestial aspiration of his intellect and the downward gravitation of his fleshly nature, even as the world soul and natural law maintain stability in the
cosmos.[70] In relation to the experience of man, the concept of the
"sacrament" takes on an extraordinary breadth, seemingly reflecting a
tension between the cosmic sacramentalism of the pseudo-Dionysius

and the conventional historical view of salvation. Though sacramental dispensations became more pronounced and efficacious as they approached the time of Christ's birth, and have, of course, assumed an absolute significance since the Resurrection, they have existed "ab initio," manifest naturally as well as in the written law and in prophecy.[71] Before and apart from Moses there were "antiqui justi" imbued with the power to do God's will, responding through the "scintilla rationis," the lingering vestige of man's original rationality, to the "umbra veritatis" which is the first and faintest of the showings forth which culminate in the incarnation of the "corpus veritatis" in the earthly life of Christ.[72] This large view of revelation has important affinities with the new interest in the experience and capabilities of natural man which appears in the twelfth century, and in the context of Hugh's thought it corresponds to a tendency to view the "opus restaurationis," the effect of grace, as precisely a restoration of the original condition of man.[73] For Hugh this is a mystical idea, the recreation in man's soul of its original perfect reflection of the divine wisdom,[74] but its articulation frequently recalls the terms in which Boethius, Firmicus Maternus, and others had extolled philosophy and the Arts, and with the concept of the *integumentum* and Plato's cosmology as an ordering framework, this sacramental process could be represented metaphorically in terms of the relations between man and nature.

In this conception, cosmology and history become intimately involved, for the ultimate object of sacramental transformation becomes precisely the original perfection implicit in the Timaean view of man, the sense of a linear, historical evolution of human history tends to merge with an emanationist conception, inspired by the thought of the pseudo-Dionysius, and more immediately by Eriugena, in which spiritual illumination is a stage in a process which begins with the procession of all life from God, and which will, inevitably it seems, culminate in the return of all life to is source. The conception of human nature which results from this larger view of life tends to treat grace, in Chenu's words, "comme une nature," [75] present in man from the beginning, and preserving in potentiality his original perfect consciousness. There is an important and impeccably orthodox precedent for this view in Augustine's trinitarian conception of the human mind, and his explanation of self-discovery as simultaneously a discovery of the Word,[76] but the strongly neo-Platonic character of the pseudo-Dionysian writings and Eriugena's system tend to obscure the role of grace, and in particular to deemphasize the pivotal significance of the unique historical incarnation in the process of illumination.[77] The result is that such obvious analogies as those between material and spiritual chaos, or between the Platonic myth of the soul's heavenly preexistence and the Christian conception of man's condition at his creation, become almost unavoidable.[78] As we will see, much of the deeper meaning of Bernardus' allegory derives from his construction of a sustained *integumentum* out of the "mystified" neo-Platonism of his late-classical sources, where the line between philosophical and religious truth is frequently almost invisible.

Perhaps the single most important point on which these "Platonisms" converge, and at the same time that which serves best to emphasize the essential differences between them, is their common concern with the *experience* of the contemplative mind in confrontation with the "liber naturae." For Hugh of St. Victor, commenting on *Ecclesiastes*,[79] the universe is finally a prison, incapable of making truly manifest the reality of God. For an anonymous disciple of Guillaume de Conches, on the other hand, a recognition of the omnipresence of the *anima mundi* is sufficient in itself to lead us "ad mentis divine cognicionem." [80] Bernardus, deeply versed in rationalist Platonism, is at the same time acutely aware of the limits of rationalism, and one of the major themes of the *Cosmographia* is the complexity, the ever-present danger of self-deception and spiritual doubt, with which the quest for a spiritual meaning in nature is fraught. Though his allegory must, I think, finally be understood as an affirmation of the sacramental reality underlying the workings of the natural order, this reality is never directly manifested in the course of the poem's action. Our view of events is confined, so to speak, by the limits of the Platonic universe; we are made to experience the struggle of natural powers to give expression to a supernatural reality, and the struggle often appears to be in vain.

In giving allegorical expression to the drama of the quest for enlightenment, Bernardus is largely developing certain hints in the

works of his own principal literary *auctores*. The *De consolatione* of
Boethius is not simply an exposition of the arguments of Philosophy,
but a dramatization of the prisoner's response to them; the cosmological imagery of the work points up the storms of doubt, the struggle of
light and darkness in his mind;[81] the allusions to Orpheus and Hercules stress the effort of withdrawing from thoughts of earthly love
and loss to a recognition of higher values. Martianus Capella's elaborate staging of the union of knowledge and eloquence includes a
careful delineation of the uncertainty, the searching by trial and error
which is a necessary prelude to this union. The *Timaeus* itself is rich
in psychological nuance, and the delicate balance of the elements, the
tension between the movements of reason and unreason in the universe anticipates the uncertain condition of the human spirit involved
in the flux of created life. The *experience* of philosophy, in short, as
well as its objects and its value in itself, is a significant element in
these works; we will see a decisive formulation of this aspect of their
meaning in the commentaries attributed to Bernardus, and in the
*Cosmographia* it is dramatized in the wanderings, the hopes and fears
of Bernardus' protagonist, the goddess Nature. For in Bernardus' allegory this figure serves not only as the principle of generation and substance, but as an index to the possibility of moral and psychological
stability in human life. She is a figure of importance and dignity: the
"I, Natura, sequar" of Bernardus' Urania, celestial principle of human
understanding, gives lofty expression to her role.[82] But her limited
power to govern and coordinate those human faculties and passions
for which she provides a pattern of potential harmony is also made
plain in Bernardus' account of her office. This same Nature thus
serves as a standard for interpreting the history of man's fall and redemption, the tension between order and violence in the course of
civilization, the complex interplay of psychological forces in all aspects of human love. In Bernardus' hands the theme of the struggle of
man to maintain right relations with the natural order, to derive enlightenment and self-knowledge from the contemplation of the universe, becomes an epic theme.

4.

*Bernardus Silvestris and His Works*

Most of what we know about Bernardus' life is based on the evidence of his relations with others.[83] The *Cosmographia* is dedicated to Thierry of Chartres; the rhetorician Matthew of Vendome was Bernardus' student, probably at Tours; he was evidently in close touch with the work of the Spanish school of translators and scientists.[84] Only two dates in his career can be even tentatively established: the *Cosmographia* was read before Pope Eugene III in 1147,[85] and the *Experimentarius* has been dated by its editor about thirty years later.[86] Several references in the former work suggest the author's fondness for Tours and its environs, and it has commonly been assumed that he studied, and perhaps taught, at Chartres. Though known mainly by his poems, to judge from the evidence of citations in the work of twelfth-century authors, he was hailed by Matthew of Vendome as "the glory of Tours, the gem of scholarship, the pride of the schools," and is commemorated by Henri d'Andeli in his *Bataille des vii ars*,[87] as

> . . . Bernardin li Sauvages
> qui connoissoit toz les langages
> des esciences et des arts.

The corpus of Bernardus' works is equally difficult to establish with certainty. The *Mathematicus* and the *Cosmographia* are coupled with his name in numerous twelfth-century citations, and the attribution of the *Experimentarius* seems equally certain, but the commentaries on Vergil and Martianus, though plainly by a single author, are assigned to Bernardus only on the evidence of a late manuscript of the former, though there is also considerable internal evidence, and the attribution has been generally accepted.[88] The commentary on Martianus makes frequent reference to still another commentary, on the *Timaeus*, which remains undiscovered.[89] Finally, the rhetoricians Eberhard the German and Gervais of Melkley mention Bernardus as having written at length on the *ars dictaminis*, though no known treatise on the subject can be positively identified as his.[90]

Even on the basis of this fragmentary evidence, however, a number of conclusions can be drawn regarding Bernardus' place in twelfth-century intellectual life. He was clearly familiar with the work of Thierry and Guillaume de Conches, and as Silverstein and Gregory have shown, the *Cosmographia* is a brilliant distillation of and comment upon complex ideas current among the Chartrians. He was also in close touch with newly circulating ideas drawn from Arab thought, and can, I think, be shown to have known at first hand the works of the great precursor of twelfth-century Platonism, Johannes Scotus Eriugena. It is equally clear from his relations with Matthew of Vendome and from the impression of his activity that emerges in the rhetorical and poetical treatises of the later twelfth century, that at least a large portion of his career was spent in the more strictly literary milieu of such schools as that at Tours.

This involvement with two spheres of activity accounts for much that is unique in Bernardus' work, and may also serve to illustrate the fortunes of Chartrian thought in the later twelfth century. The importance of Chartres itself seems to have declined rather abruptly around the mid-century, partly under the pressure of criticism from Victorine and Cistercian traditionalists, partly in the face of an increasing tendency towards specialization in the schools. Chartrian metaphysics becomes the province of those theologians inspired by Gilbert de la Porrée; the sciences of the *quadrivium*, increasingly dominated by a new awareness of Aristotle, tend to separate into individual disciplines, and the emerging universities become the centers of scientific study; the study of the *auctores*, in the meantime, becomes increasingly bound up with training in literary expression and the commentaries of the later twelfth century assume the character of handbooks for *litterati:* in place of the glosses of Guillaume de Conches on Plato and Boethius we find the mythographical compendium of Albericus of London and the *Allegoriae super Metamorphosin* of Arnulf of Orleans. In short, the Platonist synthesis of learning and expression as a means of philosophical and religious understanding had lost its hold on the minds of serious scientists and theologians.[91] To the extent that it survives in Bernardus' commentaries he may be regarded as the last of the "Chartrians," and at the same time it was through his poems that a new version of the old ideal survived as a central theme in allegorical poetry, even as his commentaries helped to effect the transition from a primarily philosophical to a primarily literary emphasis in the area of the study of the *auctores*.

Viewed from the standpoint of the development of medieval poetry this transition has, of course, more positive implications. For the range and emphasis of Bernardus' work are an assertion of the new importance of literature itself in the twelfth century, and of the willingness of poets, as well as commentators and teachers, to accept the challenge of the great *auctores*. There is no medievel precedent for Bernardus' ambitious choice of the theme and form of his *Cosmographia*. The use of the visionary dialogue, and the evocation of the personae of Philosophy and the Arts, serve a simply expository purpose in the *De eodem* of Adelhard of Bath, but Bernardus is consciously addressing what he takes to be the central theme of the greatest poetry. Though a remarkably complete rendering of the principal ideas of the Chartrians, the *Cosmographia* uses philosophy for its own purpose, exploiting to the full the implications of the ideas of integrity and order in nature which his contemporaries were formulating, subordinating them to an aesthetic design in which their sacramental analogies are subtly suggested, and in which at the same time the limitations of the natural order as a guide and stay for human life are made plain.[92]

I have ventured to call the *Cosmographia* an epic; this aspect of the work and its relation to Bernardus' special perspective on his philosophical theme may be understood more clearly when set in relation to the view of human experience which emerges in the commentaries on Vergil and Martianus. The *Aeneid*, for Bernardus, is the story of "what the human soul, placed for a time in a human body, achieves and undergoes." [93] More specifically it is the story of Aeneas' progress from infancy to maturity, from ignorance to understanding, from willfully oblivious involvement with Dido to reunion with Anchises, and a revelation of the source and end of life. Bernardus dwells at length on Aeneas' encounter with the Sibyl, his entry into the temple of Apollo, and the subsequent descent to the underworld. These represent his introduction to philosophy, and his experience in Hades is compared to that of Perseus, Theseus, and Hercules, all taken as emblems of the power of intellect to conquer the monsters of vice and ignorance.[94] The commentary breaks off just before Aeneas' encounter with Anchises, but it is clear that the latter's discourse on the fiery *spiritus* which pervades all life, and on the experience and destiny of the soul, is conceived as the reward of Aeneas' long labor.

The *Aeneid* commentary contains only the briefest of references to complex philosophical ideas, and its primary emphasis is on glossing Vergil's imagery. The commentary on Martianus, on the other hand, incorporates detailed discussions of the nature of the elements, the *anima mundi*, the origin of the human soul, the classification of the sciences, and other such technical concerns,[95] while at the same time tracing the progress of Mercury, the capacity for expression as yet unenlightened by wisdom, toward the encounter with Apollo which will enable him to transcend his ignorant condition.[96] Like the commentary on the *Aeneid*, the commentary on Martianus breaks off just at the point of a culminating revelation, in this case the manifestation of the mythological "trinity" of Jove, Pallas, and Juno, whom, says Bernardus, philosophers call "Father," "Noys," and "World Soul," and whom Scripture identifies as Father, Son, and Holy Spirit.[97] The incorporating of the substance of philosophical investigation into the context of an interpretation of Martianus' allegory is evidently intended to confirm the integrity of the imagery of the *De nuptiis*, its necesary fidelity to the Platonic archetype, and although the philosophical and the literary portions of the commentary remain largely separate, the intention is significant in itself. So Boethius is guided by Philosophy to an appreciation of God's majesty through philosophical consideration of his handiwork, as vividly realized in Platonic terms in the "O qui perpetua." And so, in the *Aeneid* commentary, Aeneas is presented as passing through education from visible to invisible things, "per creaturas ad creatorem." Bernardus associates the three *fabulae* in the prologue to the commentary on Martianus, by way of explaining the notion, at first sight rather startling, that the *De nuptiis* is an imitation of the poem of Vergil:[98]

The author's purpose is imitation, for he takes Vergil as his model. For just as in that poet's work Aeneas is led through the underworld, attended by the Sibyl, to meet Anchises, so Mercury here traverses the universe attended by Virtue to reach the court of Jove. So also in the book *De consolatione* Boethius ascends through false goods to the *summum bonum* guided by Philosophy. Thus these three *figurae* express virtually the same thing. Martianus, then, imitates Vergil, and Boethius Martianus.

To conceive of the study of the Liberal Arts as an epic theme inevitably seems a pedagogical contrivance. Bernardus' exposition is a cumbersome approach to an ideal which even Martianus, for all his pedantry, had treated more lightly. But the "experience" of philosophy is, as I have suggested, a real concern of both Boethius and Martianus, and one of which Bernardus' commentary on Martianus, by comparison with that on the *Aeneid,* reflects a deepening appreciation. For there are a number of strikingly original mythographical glosses in this commentary which, elaborating on hints in Martianus' text, do indeed present the human condition in terms that are both philosophical and heroic. Thus the phrase "joined by a sacred bond" in the hymn to Hymen which opens the *De nuptiis* suggests to Bernardus not only the role of "marriage" in the cosmos, the power which "preserves union in discord by its divine embrace," but the continuity by which life survives the recurring fact of death, and mortality is integrated with eternity. The image he finds appropriate to this idea is the sacrifice of Pollux, who bestowed a portion of his divine immortality on his mortal brother Castor: [99] "The god underwent mortal death that he might confer his godhead upon mortality; for spirit dies temporally that flesh may live eternally." The language, and the gratuitous introduction of the myth of Castor and Pollux into a context where no reference to them can be detected, are suggestive in various ways. Most obviously, there seems to be a hinted comparison between the relationship of the two brothers and the relationship of Christ's sacrifice to the mortality of mankind, an analogy comparable to, but far bolder than any suggested by Hugh of St. Victor in his sacramental analysis of the cosmogony Viewed in strictly human terms the transcending of the barrier of death through sacrifice suggests the powerful account of human reproduction as a struggle against the threat of annihilation, and the heroism of those twin brothers who battle the Fates with "generative weapons," which is the climax of the *Cosmographia.*

Another passage which stimulated the commentator to original thought is Martianus' account of the gifts bestowed upon Psyche, the

human soul, by the gods.[100] These include immortality, conferred by
Jupiter, and the "mirror of Urania," which bestows self-knowledge
and is the gift of Sophia. Venus and Vulcan give contrasting qualities:
Venus arouses in her the "itch of lust" and the desire for gratification
of the senses; but Vulcan instills in her "unquenchably enduring fires,
lest she be overwhelmed by shadow and dark night." The role of Vulcan here inspires the commentator to a long analysis of his mythical
history, conceived as a dramatization of the life of the *animus* or rational soul, and in particular of the faculty of *ingenium* within this
soul. He is innately capable of possessing wisdom, and hence sought
the love of Pallas, but he is lame, and is inseparably joined with
Venus; wisdom inevitably eludes his grasp, and he is left with the responsibility of fathoming his own complex nature as a necessary condition of proceeding to higher things.[101] Again we are presented with
an image of conflict, the struggle of an intrinsic power of mind
against the threat of spiritual blindness. Vulcan, seeking to resist the
encroachment of sensuality, is a vital link between the present state
of man and his original condition, in which he was immortal, rational, fully possessed of self-knowledge, and hence of wisdom. In
Vulcan is embodied a conception of the human situation, actual and
potential, which we will encounter again in the "genius" figures of the
*Cosmographia*.

Though the internal evidence is compelling, and will be illustrated
in more detail below, we cannot assume Bernardus' authorship of
these commentaries, and hence must not exaggerate their relevance to
the *Cosmographia*. Many of the specific glosses, notably those associated with Aeneas' descent to the underworld, are only elaborations
on glosses of Guillaume de Conches on Boethius.[102] His conception of
Vulcan is anticipated by Eriugena and Remigius of Auxerre, and Remigius and Fulgentius had touched upon the historical implications
of Venus' gifts to man and their effects on his will and consciousness.[103] But the combination of bold originality with clarity and consistency of purpose which the commentaries on Vergil and Martianus
exhibit in common with the *Cosmographia* is sufficient, whatever
their actual relation, to make comparison between them worthwhile.

More important, the *Cosmographia* incorporates and synthesizes
the different "epic" themes brought together in the commentaries. In

it the condition of man is shown by anticipation as demanding that he learn, like Boethius' prisoner, and like the hero of Bernardus' own allegorized *Aeneid*, to withstand the storms of confusion, rise above the clouds of ignorance, and so gain the reward offered by Martianus' Urania, self-knowledge combined with an understanding of the nature of cosmic and ideal reality. A long chain of fate and history depends from the first man, as Bernardus' Nature beholds him in the Table of Destiny, and there is, as in the world of Aeneas, a significant historical dimension to the universe of the *Cosmographia*. The challenge of destiny and the burden of experience seem to be inseminated in human life from the very beginning, and the conflict between order and violence in the history of man is foretold in the stars. From Nature's opening appeal to God to the final account of man's own procreative resources the poem is an image of the tension between the will to order, fulfillment, and enlightenment, on one hand, and the menace of chaos and irrationality on the other.

One further thematic comparison may be made between the *Cosmographia* and Bernardus' other writings, with regard to the role of self-realization in the poem. This theme brings into relation the concern with self-discovery implicit in Bernardus' treatment of the "epics" of his precusors Martianus, Boethius, and Vergil, and a concern with such notions as fate, astral determinism, and the scope of human freedom which, apparent in the *Experimentarius*, are dramatized at length in the *Mathematicus*. This curious poem is based on a rhetorical "dilemma." A son is born to the wife of a Roman knight, but an astrologer has foretold that he will slay his father. Having been given the cryptic name Patricida the boy is raised in secret, unbeknownst to his father, who had ordered that he be slain at birth. He grows up perfect in all respects and becomes ruler of Rome. His mother in due course reveals the truth to his father, who professes a willingness to accept the fate foretold for him, but asks that he may first meet with his son. After an emotional meeting the son, refusing to accept the tyranny of fate, convokes the Roman senate, and having elicited their promise to grant his request, announces his intention to commit suicide. The senate falls into indignant confusion until Patricida, after rebuking them, announces his abdication as ruler, and claims the freedom to choose his fate. At this point the poem ends abruptly, but it seems clear enough that Patricida must be understood to go through with his intended suicide.[104]

What gives significance to Patricida's decision is the basis on which it is made. In the declamation on which the *Mathematicus* is based, the hero acts only to spare his father, and fears, even then, that his action will cause his father to die of grief. But Bernardus' Patricida asserts his dignity as a man and the affinity of his understanding with heavenly powers, as arguments against acquiescence in the working out of an arbitrary fate. To one who understands the nature of man, he argues, death is a release, a rejection of the shadowy world of fleshly existence and a return to the heavens. Bernardus stresses the clarity of Patricida's vision by contrasting it with the determinist attitude of his father, for whom "omnia lege meant," and who sees death only in terms of "chaos and the dark realm of Stygian Jove." Again, the elaborate syllogistic arguments and rhetorical protestations with which the senate seeks to oppose Patricida's resolve are scornfully contrasted by Patricida himself with the conduct of the Romans of old, whose values assume by contrast the status of primordial simplicity and integrity.

The point of view embodied in Patricida is largely derived, I think, from an ancient astrological treatise, the *Mathesis* of Firmicus Maternus. Early in the opening book of that work the author entertains the objections of a hypothetical enemy of astrology. Why, he is asked, should a man labor to instill virtue in his character? Why should he learn to scorn death and acquire philosophy if, as the astrologers claim, all is ordained by fate and determined by the motions of the stars?[105] Patricida employs a similar antideterminist argument: "Why," he asks, "is our mind so closely aligned with heavenly powers, if it must suffer the grim necessity of harsh Lachesis? It is in vain that we possess a portion of divine understanding if our reason is unable to provide for itself."[106] Firmicus Maternus' reply to his opponent consists in turning his argument against him. The very discoveries of the *mathematici*, far from enslaving human life, confirm its transcendent origins. It is precisely when we have gained control of our sensual natures and learned to scorn death that we are able to achieve astrological learning, and what it teaches us is that we can master nature, transcend the necessity of the stars. Our slavery is not due to fate, but to our ignorance of the limits of fate. "To whom," he protests, "does the whole nature of divine science reveal itself except to one whose mind, sprung from heavenly fire, is sent forth to ensure the government of earthly frailty?" [107] Thus self-knowledge and an appreciation of the complexity and consistency of the workings of the heavens go hand in hand; similarly Patricida sees death as enabling man to regain the lofty existence of those stars with whose nature his own exhibits such affinities, and Noys, in the *Cosmographia*, proclaims that man, having learned to know the secret causes of things, and gained that mastery over nature which is his birthright, will go forth at the dissolution of his earthly life: "He will ascend the heavens, no longer an unacknowledged guest, to assume the place assigned him among the stars." [108]

The words echo Ovid's prophecy of the immortality of Augustus, and mark the culmination of what may be called the optimistic reading of human history as presented in the *Cosmographia*. Man's destiny is glorious, his very capacity for knowledge and action seem to proclaim his ultimate divinity and for a moment we acquiesce in the vision, as in the lofty prophecy of Ovid's Jupiter, or Anchises' declaration in the Vergilian Elysium that under Augustus the Golden Age will be restored. But it remains clear in the larger context of the *Cosmographia* that this ideal view of human dignity can be maintained only with difficulty in the face of what we know from experience of human life. Like the ambiguous panorama of the *Metamorphoses* and the seven years of loss and hardship which lie behind Aeneas' brief sojourn in Elysium, the nature of life as presented in the *Cosmographia* makes it hard to feel confident of man's destiny. Like the story of Patricida, that of the creation of man is left in suspense. Rich in suggestion, the ideal of self-awareness remains only another aspect of the heroic theme of the *Cosmographia*, a powerful but indecisive assertion of the dignity of man.

Moreover the *Cosmographia* is a complex treatment of philosophical ideas as well as a fabulous account of human destiny. The rather simplistic opposition of flesh and spirit on the basis of which Noys and Urania, in common with Patricida, present their promise of glory does not adequately reflect Bernardus' treatment of human consciousness in the work. Cognition and aspiration are both presented in con-

crete, psycho-physical terms as expressions of the general *motus,* the *vitalis calor* which pervades and embraces all life. Urania and Endelechia, through whom, in the terms of Bernardus' cosmic scheme, this *motus* is instilled in man, are forced to work against the dullness and intractibility of material nature, but in itself their operation is inseparable from this nature, and is contained with it by the all-embracing *ratio* of creation as it issues from the mind of Noys. Insofar as it is truly transcendent, the vision of man is owed to something beyond these cosmic powers, anterior to the primary creative tension of flesh and spirit, a *motus* which embraces all *ratio* and all physicality and which has its beginning and end in God.

## 5.
## The Sources of the Cosmographia

Specific instances of Bernardus' borrowing from earlier authors will be found in the section of this introduction dealing with the dramatis personae of the *Cosmographia* and in the notes to the translation itself. Here I intend simply to note the general character of Bernardus' use of his classical and medieval sources, since this sheds considerable light on his artistic purpose, and to characterize his use of two of the most important of these. It has been observed that the *Cosmographia* is "somewhat older-fashioned" in its affiliations than the more strictly cosmological writings of Bernardus' contemporaries,[109] but while my own investigations have largely confirmed this view, it must be carefully qualified. Bernardus is a true Chartrian in his reverence for the *auctores,* and this, together with his consciously classicizing conception of the form and theme of the *Cosmographia,* is plainly reflected in countless details of imagery and language as well as certain basic ideas. But the borrowings from medieval authors, though fewer and less obvious, are ultimately the most important in conveying the underlying implications of Bernardus' cosmogony.

After the *Timaeus* itself, which of course defined the structure and major cosmological themes of the *Cosmographia,* the most frequently used sources by far are Calcidius and the pseudo-Apuleian *Asclepius,* followed by Macrobius, Martianus Capella, and Boethius. A surprisingly large percentage of what seem bold and modern notions in Bernardus' discussion of cosmic life can be traced to these authors and other ancient texts such as the *Mathesis* of Firmicus Maternus and the Latin version of the pseudo-Aristotelian *De mundo*. Indeed, they constitute a stablizing presence relative to Bernardus' borrowings from the philosophers of his own day, and it is in relation to the large pattern of traditional associations which they provide that the more daring implications of the ideas with which he deals must be understood. From Calcidius came the conception and even the imagery of Bernardus' characterization of Silva, the primordial chaos, and he provides as well the most illuminating glosses on figures so elusive as Noys and Endelechia. Macrobius is largely responsible for such vivid details as Bernardus' use of the motif of the descent into Hades, and his apostrophe to the sun, as well as many details of the universe of the *Cosmographia*, and the emanationist character of the "Golden Chain" of life within it. Martianus' language, in particular the tone of the numerous prayers and invocations of the *De nuptiis*, pervades the *Cosmographia*, and the experience of Bernardus' Nature owes much to that of Martianus' Mercury and Philology in the course of their preparation for marriage, while the *Consolatio* of Boethius provides a running commentary on the interplay of light and shadow, turbulence and tranquillity, and its foreshadowing of the struggles of the human psyche. The account of fate and providence in the fourth book of the *De consolatione* is also an essential gloss on the relations of Bernardus' Noys, Nature, Endelechia, and Imarmene.

The contribution of the *auctores* to Bernardus' drama is thus both literary and philosophical, and they are frequently quoted verbatim, thus providing us with a useful insight into Bernardus' method of composition. Much of the meaning of the *Cosmographia* derives from his elaboration of the nuances of their imagery and diction, and his juxtaposition of their more provocative ideas in such a way that their suggestiveness is enhanced. As Silverstein has shown, nothing is more characteristic of Bernardus' philosophizing than its obliquity. He flirts continually with materialism, pantheism, Manichaeism, and a sort of cosmic eroticism, but the truths he insinuates can never be precisely defined in terms of any of these positions, and the relationship among them remains a matter of poetic association rather than rational argumentation. Reminiscences of Vergil, Ovid, and Claudian scattered through the verse sections of the work enrich its texture, and the works of these poets fundamentally influenced its larger structure, but in relation to literary tradition the *Cosmographia* is above all an anthology of major motifs from the authors in the Chartrian canon. Bernardus' continual dialogue with these authors both illumines the archetypal quality of their works and enables his own allegory to partake of that quality.

In relation to this oblique, literary handling of the *auctores* the *Asclepius* demands special consideration. This pseudohermetic work is the direct source of many of Bernardus' most challenging ideas and often of the very words in which they are presented, and provides the basis for a surprising number of the characteristically "Chartrian" features of Bernardus' work. Its treatments of the relationship of macrocosm and microcosm, genus and species, the elements and the *causae* operative in created life, virtually define the major concerns of Chartrian thought.[110] Its hints of emanationism and its emphasis on the potency instilled in earthly nature help to account for two of the most striking features of Bernardus' cosmogony.[111] Most significant of all these correspondences is its treatment of philosophical study and the contemplation of the universe as an antidote to the debilitating *malitia* which has established itself in the human constitution. Once men were capable of seeing the world whole, and as an expression of the goodness of God. But the world has drawn men's love away from God; the "vera philosophia" of contemplation and grateful prayer has been broken up into disciplines whose coherence is imperfectly understood and liable to the further contamination of sophistry.[112] Thus, Trismegistus tells his disciple Asclepius, we can only pray that God will free us from the bonds of our mortality and restore to us that visionary power through which man becomes virtually cooperant with God, completing his handiwork through technology and rational conduct, and confirming its goodness by prayer.[113]

An important complement to these themes of the *Asclepius* as they function in Bernardus' allegory is the account of the origin and descent of the human soul in Macrobius' commentary on the *Somnium Scipionis* of Cicero.[114] He describes the soul's original likeness to its creator and its necessary "degeneration" into bodily existence, where, however, it retains an essential spark of pure reason, the "particle of the divine mind" claimed by Bernardus' Patricida. This last is essential to Macrobius' scheme (as it is to Bernardus'), for his conception of the dignity of man and the purpose of life depends on the soul's immortality and the conviction that every soul must finally return to its original home in heaven.[115]

The conception of the human condition presented by Macrobius and the *Asclepius* is essentially that which evokes the promise of Urania and Noys that man will, through knowledge, reattain his original dignity and so become immortal.[116] It constitutes a fabulous counterpart to the Christian conception of fall and redemption and so reveals Bernardus' Chartrian training in an important way. For the evocative figures of the *Asclepius* provide the necessary poetic ingredient which makes of the Platonic universe an *integumentum* capable of conveying, with Macrobian decorum, the Christian doctrine which underlies it. Bernardus' allegorical world exists in a carefully defined but systematically oblique relation to sacramental reality, and to trace the sources out of which the verbal and formal patterns of the *Cosmographia* are confected is to discover a process which is the counterpart in poetic practice to the critical method of the glosses of Guillaume de Conches and Bernardus himself. And the *Cosmographia* cannot be understood without a recognition of the sustained dialectical interplay among the various sources of its traditional Platonist form and argument on one hand and, on the other, between their Platonism and the Christian neo-Platonism with which this structure is, as it were, informed.

The most important medieval source of this Christian neo-Platonist element in the *Cosmographia* seems to me to be the *De divisione naturae* of Johannes Scotus Eriugena, though Bernardus may possibly have received Scotist doctrines indirectly through some such medium as the *Clavis physicae* of Honorius Augustodunensis.[117] It is almost certainly Eriugena who provides the theological and philosophical basis for the dynamic view of material potentiality which is so important a feature of Bernardus' allegory, and there are richly suggestive anticipations in the *De divisione* of such elusive topics as the relation between human reason and God's wisdom, between Urania and Noys in the terms of the *Cosmographia*. It is in connection with such complex considerations, as we will see, that Bernardus points up the limitations of the strictly cosmological conception of man, and the final inadequacy of the natural world as a source of vision and stability; and when reinforced by the bold and sweeping neo-Platonist vision of Eriugena Bernardus' conception of these limitations assumes a positive as well as a negative aspect. Though documentation is particularly difficult, since the special decorum of Bernardus' mythic narrative allowed only the most oblique expression of spiritual themes, I have discovered no single work which corresponds so closely to Bernardus' conception of the human situation as the *De divisione*.

Bernardus' borrowings from contemporaries are of a more miscellaneous character. The *De vi dierum operibus* of Thierry of Chartres undoubtedly contributed to Bernardus' description of the work of Noys, and what seems like a radical emanationism in his treatment of the capacities of matter very probably reflects, in addition to the influence of Eriugena, an elaboration of certain hints in Thierry's analysis of the relations of matter and form. Certain words and certain details of his account of the *ligamina* joining heavenly and earthly life seem to reflect the influence of the Latin versions of the *Introductorium maius* of Abu Ma'shar. Numerous points of detail in his account of the relations of heaven and earth can be paralleled in the writings of Hermann of Carinthia and Dominicus Gundisalvus, and the commentaries of Guillaume de Conches and the medical works translated and adapted by Constantinus Africanus provide useful glosses on his descriptions of the elements.

But these concrete borrowings, while they confirm Bernardus' close knowledge of contemporary thought, do not convey its full importance for the *Cosmographia*. This is more a matter of Bernardus' complex perspective on the Platonisms with which he deals. He is true to the scientific spirit of his day in his rigorous de-mystifying of the goddesses who conduct the action of the *Cosmographia;* Noys and Endelechia are clearly subordinated to the true godhead, and wholly concerned with the affairs of the universe. And as these figures become vividly realized presences in Bernardus' poetic world his sense of their operation in physical nature becomes correspondingly sharp. Physis works in response to the intuitions of Theory, and thus brings the practical aspects of man's relations with Nature into conformity with the ideal pattern which her daughter discerns, just as Urania seeks to discipline his mind. Thus various kinds of twelfth-century scientific thought, physical and technical, conspire to produce a universe so unified in itself as to defy any facile idealist or hierarchical conception of its relation to God. And it is this self-contained structure which provides a foil to the Christian neo-Platonism through which Bernardus' sense of man's more profound relationship with the universe is conveyed.

But as has already been suggested, the sources and precise definition of the ideas with which Bernardus deals are fully meaningful only when viewed as part of the complete machinery of the *Cosmographia.* In the following two sections I will try to suggest how Bernardus brings these ideas to life, first as incarnate in the goddesses and "genii" whose relations define the action of the work, and then as determining factors in the form and structure of the allegory itself.

## 6.

## Dramatis Personae

Nature, Noys, and Silva are the most important figures in the scheme of Bernardus' allegory, and their relation must be clearly understood. For though the contingency of all *naturalia* is perfectly clear in the scheme of Bernardus' cosmogony, he is at pains to endow natural life with as much initiative as possible. Hence the undisciplined passion of Silva, as transmitted by Nature, seems to provide the impetus which draws Noys forth from eternal deliberation into action, and it becomes important to determine as precisely as possible the degree of autonomy assigned to each figure.

Nature is Noys' daughter, "filia Providentiae," ceaselessly and ardently concerned with the well-being of material life.[118] In the first book of the *Cosmographia,* dealing with the creation of the universe at large, she has little to do once Noys has been invoked, and simply stands by while her domain is defined by Noys and animated by Endelechia. Nevertheless, by the end of the final section of Book One, a vivid account of the life of the universe, Nature has taken her place among the cosmic powers: "Natura elementans," identified with the firmament and the stars, is cited as the power through which the elements are brought forth from Silva, and "Natura artifex" is described as providing dwelling-places for the souls which Endelechia transmits to her at the bidding of Noys.[119] In Book Two she becomes more active, summoning Urania and Physis to the task of providing a soul and body for man, and herself ensuring the "formative uniting" of the two principles "through emulation of the order of the heavens." [120] Her role is suggested also by the response of earthly nature to her appearance in the earthly Paradise: the earth erupts in an orgy of fertility, "for it had received a premonition that Nature, mother of generation, was at hand." [121]

Nature's role is thus one of containment and coordination rather than creation. She administers the initial impression of matter with elemental form, mediates between the animating power of Endelechia and the orderly processes of generation and perpetuation associated with Imarmene, and in an analogous manner aligns the celestial principle with the earthly in the constitution of man. Thus, though she cannot be identified with any such active principle as the *ignis artifex*, she perfectly embodies the twelfth-century conception of a "legitimum societatis foedus" between heaven and earth, embracing a complex of harmonious and autonomous causes. "The nature of the universe," says Bernardus, "outlives itself, for it flows back into itself, and so survives and is nourished by its very flowing away." [122] And this is a fair reflection of the qualities for which Nature stands: order, perpetuity, continuity, unflagging vitality.

As the "womb" of life, the vessel within which heavenly principle and earthly substance come together, Nature embraces the same gamut of being as man himself, and thus serves as man's surrogate in the action of the *Cosmographia*. As such her role is more complex, and its significance is conveyed iconographically. Bernardus is plainly concerned to emphasize her passionate femininity in the opening scene of the poem, based on a scene in Claudian's *De raptu Proserpine*, and strongly recollective of the appeals of the heroines of Ovid's *Heroides*. The passage may also make us think of other such passionate voices: of Vergil's Dido, and especially his Venus, beseeching Jove to grant rest and a renewal of civilization to her exiled son

Aeneas and his followers; of the voice of Israel, calling through the psalmist to a God who has turned away His face. In Book Two Nature becomes a questing figure, subjected to a wearying journey through the universe, and made to contemplate a grim vignette of the dark and somber aspects of human life in the plight of the unborn souls clustered about the house of Cancer, before attaining the lofty plane where Urania waits to reassure her about the glorious destiny of her human progeny and the ultimate containment of death by life. The most obvious source for this aspect of Nature's role is the opening book of Martianus' *De nuptiis*, where Mercury, after a long search, discovers the grove of Apollo and there receives a prophecy of his marriage with Philology, the sum of all knowledge.[123] But there is a dark undertone in Bernardus' account which is closer to Vergil, and it is finally to the *Aeneid* and Aeneas' encounter with Anchises, that we must look for an adequate precedent for Nature's experience. Anchises' prophecy is to the history of Rome as Urania's proud assertion of human destiny to the labors of the "genii" of human generation; both rest in uneasy juxtaposition with a vision of life as ceaseless labor, and it is Nature, with her anxious concern for the well-being of all created life, who acts as the protagonist in this struggle.

As the eternal vitality which Nature transmits derives from a higher power, so the passion with which she invokes this power derives from her involvement with the energy and fecundity of Silva, the primordial chaos. Calcidius, paraphrasing a passage in Aristotle's *Physics*, provided Bernardus with the image of a matter which "longs for adornment just as female desires male and what is shapeless longs for beauty," [124] and on the basis of this Bernardus elaborates the splendidly flamboyant opening of his poem, where Silva is presented as weeping and yearning to be granted the impress of form and *ratio*.

Silva (or Hyle as Bernardus often calls her), though not eternal, existed before the creation, as a principle in the mind of God and as preexistent in "the spirit of eternal vitality." [125] At one point Bernardus declares that without the informing influence of the heavens, the elements sprung from Silva would be effectively lifeless,[126] and yet Urania in her difficult poetic account of the relations of life and death asserts that the essence of life is not a product of the union of form and matter, but inheres in subject matter alone,[127] thus suggesting in Silva herself, as Dronke remarks, traces of divinity.[128]

Though there is precedent in Eriugena for locating Hyle in the mind of God, conceived as archetype of the universe,[129] the aspiration Bernardus imputes to Silva is harder to assess. Calcidius uses Aristotle's image of feminine yearning, as he carefully explains, to suggest, not the sort of desire that animals feel, but rather that aptitude or potentiality that any partially formed thing ("coeptum et inchoatum") has for definitive form, in union with which "it may come to flower."[130] Thierry of Chartres, in explaining the relation of form to matter, had characterized the latter as "that possibility which alone embraces all things in itself," the state in which all creation has eternally reposed "in the simplicity of the mind of God." Hence things are brought forth, from possibility into actuality, through union with forms, but their being and individuality are owed to their preexistence in possibility.[131] This possibility or aptitude of matter seems to constitute the philosophical basis of Bernardus' more dynamic conception, and we may perhaps see in Urania's declaration that "form flows away, the essence of the thing remains" a conscious repudiation of the Boethian dictum "omne esse ex forma,"[132] commonly cited in twelfth-century discussions of the problem. It is possible to see already in Thierry's formulation a tendency to emphasize the importance of material potentiality as against that of form; other treatises of the period, and most strikingly the *De processione mundi* of Dominicus Gundisalvus,[133] use a strongly emanationist language to describe their relations. But the single most striking parallel to Bernardus' conception is provided by the *De divisione naturae* of Eriugena; here the "informitas" of inchoate matter is described, in a way which brings Calcidius' analysis into conjunction with a more spiritual view, as a "motus," an appetition seeking to pass from non-being to being, and ultimately to reunite itself with God.[134] In language which closely resembles that of Bernardus' proem he speaks of matter as "yearning to be shaped by the various 'numbers' of sensible creatures,"[135] and he anticipates the allegorizing of the creation of Eve in the *De philosophia mundi* of Guillaume de Conches when he takes God's employment of "lutum de terra" to fashion the human body as illustrating the *actio* of all created life toward its divine source.[136] Thus there are ample grounds for seeing in Bernardus' emphasis on the aspiration of matter, together with the Scotist tone of his reference to its preexistence,[137] the implication of an emanationist conception of matter as informed with a will to reascend the scale of being and reunite with God, a process to which formal existence "in corpore" would bear only a temporary and accidental relation.[138] The real animating or activating principle in matter is thus something more mysterious, a divine rather than a natural phenomenon, and provides the first of many instances in which Bernardus seems to use the Platonism of the Calcidian *Timaeus* as a foil to the far more allusive presentation of a finally spiritual doctrine.

Despite her yearning to transcend her original condition, Hyle-Silva is pervaded by an innate *malignitas*, a resistance to form and order, which can be curbed, and in the universe at large neutralized, but nowhere wholly eliminated.[139] This fact, taken in conjunction with Hyle's quasi-eternity, and other difficulties of the *Cosmographia*, suggests a kind of Manichaeanism; yet it seems clear that Bernardus is using the fact of necessity, the "errant cause," as a metaphor for the effects of original sin in human life.[140] As such it forms a part of a larger pattern to be discussed below, whereby the creation depicted in Bernardus' allegory is made to prefigure both the necessity and the fact of the restoration to come. The author of the *Asclepius* and Calcidius had recognized the *malitia* of matter as admitting a psychological interpretation,[141] and Augustine, followed by Hugh of St. Victor, had seen in the tumult of its initial "informitas" or "forma confusionis" the plight of the spirit in the absence of God.[142] Eriugena, for whom man's material body is in itself a consequence of and punishment for the sin of Adam,[143] sees material existence as involving the aspiration of an all-pervading "vitalis motus" against the hostile action of "adversae virtutes" which resist its influence.[144] Bernardus himself compares the hypothetical effects of a too strong *malignitas* in the cosmic order to the hapless state of the "imbecile" tribe of men, whose natures are incapable of maintaining an equilibrium among hostile qualities.[145] Again, this is a question which must be considered in the framework of the larger quasi-sacramental analogy between Bernardus' cosmogony and the evolution of sacred history.

The impressive figure of Noys or Providence, as Silverstein has shown, is very largely modeled on the biblical Sapientia, and the Pallas-Minerva of mythography in the tradition of Martianus.[146] She is the mother of Nature and the source, "quadam emanatione," of Endelechia, the world soul. She contains "velut in speculo" the ideas of all things, and is in some sense the "alter ego" of God. Despite the common twelfth-century tendency to identify the "archetypus mundi" which Noys clearly represents with the "Verbum Dei," Second Person of the Trinity, her precise relation to the Logos is difficult to define.[147] For the Trinity is clearly and separately described in the *Cosmographia,* and Noys seems at one point to define her role as subordinate to that conjunction of act, deliberation, and will in which the work of the Trinity consists. Noys is evidently to be understood, moreover, as inhabiting the heaven of the Thrones, third in the angelic hierarchy.[148] In view of her role in the *Cosmographia* and her description of herself as "the consummate and profound reason of God," we may perhaps distinguish her from the true "mens Dei" as a special aspect or function of this mind as it bears on the created universe, comprehending the work of Endelechia, Nature, and Imarmene and conforming them to the will of God. She is moved by prayer;[149] the brightening of her countenance and her slow, majestic response to the passionate harangue of Nature suggest that she has been brooding almost sadly over the emptiness of the precreation, and she regards Nature's vitality and zeal with the affectionate tolerance of age.

The relations of Nature, Silva, and Noys establish philosophically and in terms of mythographic allusion the limits of Bernardus' universe, and account for the gamut of experience and emotion with which his cosmogony is charged. Two less prominent figures, Endelechia and Imarmene, complete the process of creation and initiate the orderly sequence of cosmic life. The second, lifted more or less bodily from the *Asclepius,* is the principle of temporal continuity, a "serial law" in Etienne Gilson's phrase,[150] ensuring the orderly succession of growth and renewal in the universe. Endelechia, the world soul, is more complex. She issues forth from Noys by "a sort of emanation" like the biblical Sapientia, and is wedded to the world born of Silva, which she imbues in all its parts with vitality. Her power, her "fomes vivificus,"[151] derives from the heavens, and it is through the "permix-

tio" effected by the union of this power with material existence that
Nature becomes "mater generationis."

Bernardus seems to be at pains to restrict Endelechia to the status
of a cosmic power; as Noys is clearly not the "Verbum Dei," so Endelechia is clearly not to be regarded as a manifestation of the Holy
Spirit. In this connection the most valuable gloss on her role is provided by Calcidius' discussion of Aristotle's use of the term to signify
"the soul of a natural organic body," the "perfection" of material
being, that by which the possibility of matter is realized.[152] Speaking
on behalf of the "Platonici," Calcidius takes issue with Aristotle's application of the term "soul," or "anima," to what, he says, is really
only the accidental form of a body, source of its name but not of its
being, and what, moreover, is observed in things which clearly are
not animate.[153] The distinction between Endelechia and soul here
may be compared with that between form and essence in Bernardus'
treatment of Silva. The perfection which Endelechia bestows is a constant power in cosmic life, imparted to creatures in proportion as
their relative grossness will allow, and "rendered back whole and undiminished by each individual creature." [154] The identity it bestows is
ephemeral. As Urania says of form,[155] "the subject matter remains
the same, though its form pass away, and a new form only gives this
matter a new name."

This wholly cosmological conception of Endelechia reflects in a
well-developed form the influence of the Stoic conception of the
"ignis artifex," and conforms to a "physicalizing" tendency in the
twelfth-century treatment of the world soul.[156] It is reinforced by Bernardus' account in Book Two of the sun, whom Nature and Urania
encounter in the course of their descent through the heavens, and
who is described in the same sapiential language used of Endelechia
herself. The passage on the sun, after an incantatory account of its
role as "mind of the universe, the spark of perception ('fomes sensifi-
cus') in creatures, source of the power of the heavenly bodies," alludes to the sun-god's bow and lyre, and enumerates his offspring:
these include "Fruit of the Spring," "a harmless Phaethon," and "Swiftness," who seem to be functions of the sun's temporal role,[157] and also
Psyche, who is busy gathering those "igniculos" with which the sun
imbues heaven and earth.

The reasons for the presence of Psyche, the task in which she is engaged, and the suggestion, conveyed in Bernardus' language, of a sapiential affinity between the sun and Endelechia, are extremely complicated. Their common basis is Martianus' *De nuptiis*. The sapiential language which links Endelechia and the sun is drawn from Philology's prayer to the sun as she ascends through the spheres and from a hymn to Minerva later in the *De nuptiis*. Psyche is introduced in Book One of Martianus' allegory as "daughter of Endelechia and of the sun," and the "igniculi" which she is gathering suggest the unquenchable fires ("insopibiles igniculos") with which Vulcan endows her in Martianus' fable. These fires, it is worth noting, are glossed by Eriugena and Remigius as alluding to Vulcan's association with the "vitalis calor" instilled in all creation, source of vitality and generation, and, in humans, of "ingenium," the power of apprehension or imagination.[158] These "igniculi," says Eriugena, preserve in the soul a recollection of its original dignity and vision before the Fall.

The multiple associations of this image serve to crystallize the psychology which Bernardus employs in the *Cosmographia,* and define its relation to Endelechia. For the effect of Bernardus' association of Endelechia with the sun is to emphasize the tendency of all Bernardus' "goddesses" to concentrate their energies in a common Nature. As conceived in natural terms, human awareness and "ingenium" are derived from the "vitalis motus" which pervades all life and which is at once a physical and an intellectual principle ("closely akin to the atmosphere, and at the same time to the heaven itself," as Bernardus says of Endelechia). The "calor voluptatis," and the "seeds of virtue" are instilled in the human constitution from the same source, testifying to the integrity of man's original nature, in which physical and intellectual life were harmonized.[159] And in the present state of human nature the "ingenium" which man derives from this source preserves in him a dim consciousness of that spark of reason which still exists in the depths of his nature. Through this he maintains some sense of his dignity, his duties in this life, and that destiny which Noys and Urania proclaim for him in a life beyond. It has affinities with the "sacred and blessed instincts" which prompt Nature to make her initial appeal to Noys, and it is reflected in the intuitions which guide Physis in her creation of man's body. Its relation to the Platonic intimation of a prior condition of omniscience which serves Bernardus as a metaphor for man's paradisal state, and hence to a more spiritual view of knowledge, will be discussed below. For the moment it is sufficient to recognize how Bernardus' free handling of Endelechia serves to link poetically concepts of diverse origins. As the outlines of the conventional Timaean cosmology are blurred, and its doctrines mystified, by the quasi-mystical neo-Platonism of Martianus and the *Asclepius*, we begin to sense the necessary relation of all the forces of vitality and cognition to some more fundamental source.

The nature and limitations of human consciousness are further illustrated in Urania and Physis, Nature's collaborators in the creation of man. The former provides her own explanation of her role: [160]

The human soul must be guided by me through all the realms of heaven, that it may have knowledge: of the laws of the fates, and inexorable destiny, and the shiftings of unstable fortune; what occurrences are wholly open to the determination of will, what is subject to necessity, and what is subject to uncertain accident; how, by the power of memory, she may recall many of these things which she sees, being not wholly without recollection . . .

In the *De nuptiis* it is Urania, the muse of cosmic knowledge, who bestows upon Psyche the gift of self-knowledge,[161] and this is her role in Bernardus' scheme as well. Through her man's soul will be aligned with the principles of celestial order and harmony, revealing its intrinsic affinity with the divine *ratio* of creation. The implications of this affinity are conveyed by the image of the Mirror of Providence which Noys presents to Urania for guidance in the task of "the composition of a soul [for man] from Endelechia and the edifying power of the virtues." The Mirror contains the wisdom of Noys, "mens aeterna," the vision of life in its cosmic and temporal totality which man enjoyed in his original condition and the abiding capacity for which is hailed by Urania and Noys as the sign of his destiny.

Physis represents knowledge of a more immediately practical kind, the insight into the workings of elemental nature which enables man in some measure to regulate his own nature, adapting it to principles empirically deduced and experimentally mastered. There is an element of mystery in her work, an inscrutability in the design by which the disposition of the body is determined,[162] but on the whole she is characterized by her empiricism. While Urania is in general a Platonist, Physis has mastered the Aristotelian categories,[163] and it is the power to make distinctions on the basis of observation that enables her by trial and error, to devise a method of yoking the unruly elements which make up the human constitution. Historically her collaborative role in the *Cosmographia* is significant, for it reflects that union of medical study with natural philosophy in a larger sense which is one of the important achievements of twelfth-century science.[164]

Urania's enthusiastic participation in the creation of man seems to promise much, but we must recognize in the vision she bestows the same limitations that define the activity of Endelechia. The materials with which she works, indeed, are Endelechia herself, natural virtue, and a knowledge corresponding to Endelechia's hierarchical relationship with Noys. The soul she instills in man is comparably limited; its capacity for supreme knowledge is undeniable, but its operation is hindered by the effects of necessity (and, by implication, sin) in human life. Thus, while Urania's work, like that of Noys herself, is guided by the eternal exemplars of creation as they appear in her *speculum*, Nature must manage with the Table of Fate, in which truth is revealed only as it manifests itself in the temporal order, in the workings of the universe, and in recorded history, while Physis has only the Book of Memory, wherein is recorded "nothing else but the intellect applying itself to the study of creation and committing to memory its reasonings, based often upon fact, but more often upon probable conjecture." [165] The narrowing range of speculation thus defined corresponds to an increasing involvement with the turbulence and confusion of sublunar, elemental existence, and reinforces the theme of conflict and uncertainty. Like the vivifying power of Endelechia, the vision of Urania grows dimmer as the soul is submerged in the tide of existence, and the marriage of soul with the body tenuously synthesized by Physis is in itself no guarantee of the fulfillment of that destiny which Urania so proudly hails. Just as we can perceive the operation of Endelechia, but are confounded as to its source and true nature, so a recognition in principle of the integrity and intrinsic divinity of the vision of Urania does not enable man to escape necessity. Fate, and conjecture, remain his principal resources, and wisdom is at best a spark, an "igniculus" which perpetually illumines his inner mind but remains almost inaccessible to his ordinary awareness.

It is with this conception of human consciousness in mind that we must consider the roles of the various "genius" figures in the *Cosmographia*. We first encounter such a figure in the Genius or *Oyarses* of the firmament, ceaselessly engaged in inscribing forms upon all created things and thereby ensuring that they will conform to the influence of the heavens.[166] It is this Genius who introduces Nature to Urania after she has carried her long and frustrating search to the outer limits of the universe. In the course of their descent to the abode of Physis, Urania points out other genii, tutelary spirits subordinate in rank to the angels, who watch over man and subtly guide his mind "praesagio vel soporis imagine," helping it to act virtuously and avoid evil. Finally, in the concluding account of the human body, the process of generation is said to be in the charge of two genii who oversee the work of the genitalia and are also, perhaps, to be identified with the tutelary spirits of marriage.

Whatever the precise relation of these several genii among themselves, it is clear that they all exist on the periphery of consciousness: subliminally, as the unconsciously heeded guiding principles of natural virtue and procreation, or on a transcendent level, as the mediating link between Nature and Urania, natural understanding and true knowledge. As such they perform a crucial function, for they are the vehicle of such contact as there may be between man's present state and the sources of his original perfection, and ensure that his thoughts and actions do not tend to self-destruction through a total neglect of virtue and responsibility. Again we may recall the *De nuptiis* and Vulcan's gift to the soul of that unquenchable spark which keeps it from becoming wholly lost in the darkness of earthly life.[167] And it is surely to some such concept as this, in which the "vitalis calor" of physical nurture and generation and the creative and visionary "ingenium" of human consciousness meet, that we are to relate the activity of Bernardus' genii. They serve both to confirm the basic integrity of man's nature and to define the limits of his rational awareness, and so expose the limits of his power to control consciously the different elements of his constitution. The source and function of these genii is ultimately providential, but their work is performed in darkness; they communicate their influence through the stimulus of desire or imagination, and hence their effect is always liable to be offset by the random disruptions which afflict human *imbecillitas*.

The pattern in which the goddesses of the *Cosmographia* are linked serves finally to point up the crucial importance of certain relationships which, at the same time, they do not in themselves adequately define. The normative, coordinating role of Nature implies one standard of human behavior, the transcendent vision of Urania another; and it is clear that the intuitions and aspirations of Nature and Physis are inadequate to bridge the gap between these two conceptions of man's relations to the universe. Natural existence is from one point of view a means whereby man can achieve self-recognition and fulfillment; from another, it is an imprisonment, a confinement within an order which is finally an obstacle to the vision necessary for self-realization. Though there is, as we will see, a point of view from which this impasse appears in a more positive light, in terms of Bernardus' allegory it is only the genius figures that enable us to sense a continuity between the different levels of reality. Joining man's loftiest aspirations to the most fundamental impulses of his sensual nature, they complete Bernardus' presentation of human life, tenuously but perpetually suspended between transcendent illumination and the threat of a reversion to chaos.

# 7.
# The Structure and Themes of the Cosmographia

The *Cosmographia*, then, presents human life in a double perspective. We see it first of all in terms of the orderly relation of cosmic principles. All life emanates from Noys, to be preserved in structural and temporal continuity by Nature and Imarmene, and animated or vivified by Endelechia. Human life reflects the same relationships; Urania contains a version "in speculo" of the wisdom of Noys, and this is brought together with the vitality of Endelechia in a soul which Nature and virtue are to preserve in a harmonious relation with the body. But there are, as we have just seen, problems in this scheme, discontinuities in the relationship between man and the universe which present a challenge to the optimistic assumptions of "conventional" Platonism. While the Platonist vision of Noys and Urania affirms the heavenly affinity and final glory of the human spirit, we also see things from the vantage point of experience, as Nature journeys anxiously through the spheres in search of Urania, and Physis fumbles toward the proper formula for creating man's body.

The drama of the *Cosmographia* consists in its foreshadowing of the striving of a heroic human nature to break free of the constraint of ignorance and necessity and unite with the visionary perspective of Urania and unfallen man. The result we are shown is a compromise: as Vulcan sought the love of Pallas, but inevitably failed to realize this goal, spending his seed upon the earth to produce the biform Erichthonius, so Bernardus' procreative genii keep the threat of annihilation always at bay, but are never granted a respite from the labor of resistance.

One means by which this complex perspective is conveyed is the use of a wide range of tones in the different sections of the poem. Thus Book One opens with the Ovidian passion of Nature's appeal to Noys. This is followed by the long section of expository prose in which Noys patiently explains the task of creation and is then described with equal care as she performs it. The terse, lucid couplets of the still longer poem which follows depict the domain of Nature and Imarmene, suggesting the power of destiny by their very conciseness, and then, in the final section, the life of the Megacosmos is celebrated in a lush neo-Platonic prose confected out of many sources and emphasizing the vibrant and inexhaustible life of the cosmos. Book Two passes from Noys' enthusiastic review of her handiwork to the more somber tone of the description of Nature's search for Urania, and back again to idealism in the glowing rhetoric of Urania's welcoming speech. The largely expository sections which follow allude continually to the range of attitudes implied by the contrasting tones of the earlier sections; these implications, positive and negative, are crystallized metaphorically in the Mirror, the Table, and the Book which guide the goddesses in the creation of man, and in which a somber suggestion of the smallness and frailty of man darkens the presentation of the rich variety of human life and history.

There are as well subtler and more complicated ways in which the structure of the *Cosmographia* brings perspectives into focus, and establishes a clear hierarchy among what might be called the fatalist or determinist, the cosmological, and the emanationist views of human life. The first of these is most clearly expressed in the long catalogue poem which is the third section of Book One, and above all in the passage which describes the events of human history as set forth in cipher in the stars. A panorama of ancient history is presented, balancing art, learning, and the achievements of civilization against the discord and excess which tend always to destroy it, and the culmination of the passage is a couplet on the Incarnation: [168]

A tender virgin gives birth to Christ, at once the idea and the embodiment of God, and earthly existence realizes true divinity.

Much has been made of the determinism seemingly implied by the passage as a whole, and too little of the way in which this concluding couplet sets determinism in perspective. For Christ is at once the idea ("exemplar") and the embodiment ("specimen") of God, the complete embodiment of a divinity which other examples of human excellence, heroic and intellectual, manifest only partially and indirectly. His incursion into the "fatalis series" of human history bears the same relation to the general character of that history, as implied by the review of civilization which provides its context, that the "igniculus" of divine reason bears to human consciousness. How tenuous the relationship is is suggested by Bernardus' concluding comment on the astral portion of his catalogue: [169]

The stars, which the present age calls by this name or that, existed at the birth of time as heavenly fire. Lest he should stumble in seeking to express this universal theme in common speech, man created those names which even now denote the stars.

The idea of an archetypal language embodied in the patterns of the stars and transmitted intact from age to age is a noble one, but when we realize the vastness of the design it seeks to comprehend we may question the possibility of using it with confidence. The universe of meaning cryptically described in the stars is ultimately the substance of the deliberations of Noys, accessible in its full significance only to the innermost mind in the Mirror of Providence. In the Mirror itself the place of man can be determined only with difficulty, while in the Table of Fate he is all but lost among the species of created life, and civilization appears only as a progressive decline from the Golden Age to the Age of Iron. One effect of the catalogue, then, is finally to dramatize the need for inspiration, for a vision capable of rising above the proliferation of events and seeing them in their true perspective.

Viewed in this way the point of the astral catalogue is similar to that of the astrological theme of the *Mathematicus*, where the tension between the hero's conviction of man's glorious destiny and his father's acquiescence in fate expresses a great part of the poem's meaning. Whatever the intrinsic meaning of Patricida's sense of a transcendent destiny, his suicide constitutes a suspension of the course of fate comparable, in this one respect at least, to the Incarnation and Resurrection of Christ, and serving in a similar way to point up the limitations of the astrological view.[170]

The function of the astral catalogue is to a certain extent reflected in the catalogue poem as a whole, though its diffuseness makes precise definition difficult. Earthly nature, too, is made to reflect the history and character of man, his achievements and his wars, and provides a more detailed setting for the same gamut of human attitudes defined by the stars. Thus famous rivers are associated with conquest and imperial power, groves with the seclusion of poets and philosophers, mountains with the striving of human understanding to gain insight into higher things. These are interspersed with accounts of flora and fauna—the domain of Physis complementing the celestial panorama which is Urania's concern; these, while they contribute an occasional moralization, are otherwise connotatively neutral, but the cumulative effect of the catalogue as a whole is not. Aesthetically, its principle is very broadly comparable to that of such "set pieces" as the embroidery of Claudian's Proserpine which interweaves with an account of the cosmogony a prophecy of her own fate at the hands of Pluto; the panorama of history shown to Vergil's Aeneas by Anchises, and that inscribed on his shield, dramatizing the strife and glory of the Rome to come; the song of the bard Iopas, which tells of "the wandering moon and the sun's labors," images which impose a kind of necessity on the love of Aeneas and Dido. In short, the poem is an image of the course of human life, premonitory of joy and sorrow, but suspending any final judgment. The controlling function exercised by its form, diffuse and repetitive as it is, serves to create an effect of vitality and richness as well as strife. In this it recapitulates the effect of the cosmogony itself as described in the preceding sections. The cosmogony is an inspiring, and as I have suggested an heroic theme. The origins and destiny of man are almost a rallying cry for the goddesses, and Physis, confused and doubtful as she contemplates the chaotic materials out of which she is to shape the human body, has only to recall what is prophesied of mankind to feel a new surge of confidence, and enter upon her task more boldly.[171] If Bernardus' power of composition is not always equal to his conception, it is nonetheless with a very sure aesthetic sense that he lets the heroic implications of his theme create their own effect. We may even compare his use of history with that of Milton, whose Adam is shown at length the chain of events which is to follow from his original sin and is somehow emotionally purged and strengthened by the very spectacle.

The theme of human history in the *Cosmographia* is endowed with other, more explicit epic associations as well. I have suggested a general relationship between the themes of the *Cosmographia* and the reading of the *Aeneid* presented in Bernardus' commentary. The *Metamorphoses* of Ovid and Claudian's *De raptu Proserpinae* seem to have affected the very structure of Bernardus' allegory. The complex "human" experience of Nature between the completion of the cosmogony of the opening book and the final creation of man comes to a dramatic climax in the central portions of Book Two, where Urania and Noys appear and proclaim the destiny of man. Urania's account of the enduring *esse* of cosmic life, superior to death and dissolution, is modeled on the speech of Pythagoras in the final book of the *Metamorphoses*,[172] and her proclamation of man's capacity to attain immortality compliments the philosopher's "caelumque erit exitus illi." The theme of immortality is reaffirmed by Bernardus' Noys in her charge to Nature, Urania, and Physis, and Ovid's Jupiter, foretelling the deification of Augustus,[173] but in both cases this affirmation is qualified by the moral tone of the preceding discourse. For Urania and Pythagoras alike dwell on the dangers to man's spiritual well-being raised by his persistent carnality, and thus offer a broad moral and psychological perspective on the panoramic view of life offered in the two poems, at the same time that Pythagoras' doctrine of metempsychosis and the submerged neo-Platonism of Urania hint at an ultimate purgation of the ills of life.[174]

Comparison with the *De raptu Proserpinae* serves to bring certain of the darker aspects of Bernardus' cosmogony into clearer focus. An implicit theme of Claudian's unfinished epyllion is the presence in nature of a potentially chaotic *malignitas,* given expression in Pluto's threatening challenge to Jupiter, and his promise to release the forces of night upon the universe if he is not granted a bride.[175] Pluto's speech is echoed by Nature's appeal on behalf of Silva in Bernardus' proem, and thus provides a mythological counterpart to Bernardus' conception of the innate tendency of matter to resist order and life.

But Nature has a place of her own in Claudian's scheme, and she in turn appeals to Jupiter in a speech which Bernardus' proem also recalls. She urges Jupiter to counter the influence of Pluto, make the earth fruitful again, and grant to man a life worthy of his semi-divine nature.[176] The balance between Pluto and Nature, both of whose claims find expression in Bernardus' account of Silva and her evolution, defines the uncertain condition of earthly existence and thus provides a universal context for the presentation of the condition of man. That man's nature is beset by irrational and passionate forces comparable to the cosmic power of Pluto is a common theme of the two poets; Claudian's Proserpina is consistently associated with a pristine purity and innocence. Her dress is inscribed with an account of the birth and infancy of the sun and moon (surely the source of certain details in Bernardus' description of Granusion).[177] She is first shown embroidering a panorama of the cosmogony and the natural order, but her work is unfinished when she is interrupted by Venus, who is to be the "dux femina facti" in her rape.[178] The rape itself and its natural consequences are later explained by Jupiter as a challenge to human civilization, a necessary destruction of the idyll of the Golden Age, almost a fortunate fall,[179] though the unfinished state of the *De raptu* precludes any final word on the subject. In Bernardus' version, the "fall" introduces a world in which the Age of Iron is established,[180] where human achievement coalesces and dissolves, expressing the tension between a *motus* to fulfillment and a precipitate tendency to violence and self-destruction.

But while such literary parallels help to define the structure and themes of the *Cosmographia* they do not finally serve to suggest the full depth of Bernardus' exploration of the implications of the cosmic "fight against formlessness." For on the deepest level of its meaning the poem attains a perspective on the very cosmology which provides its controlling metaphor. It is no accident that Nature encounters Urania where she does, and that in order to commune with her sister she is made to pass, with the assistance of the Genius of the firmament, outside the limits of the cosmic structure; for cosmic order, which in conventional Platonic terms is the pattern of human life, all too easily becomes a measure of man's failure to conform to this pattern, and as the development of Bernardus' "epic" theme gradually forces us to adopt a historical as well as a cosmological perspective on man, the idea of cosmic harmony comes increasingly to define a barrier which man must cross in order to realize his destiny. Cosmic order, then, is to human experience as the accidental, transitory discipline of form to the aspiration of Silva, the confining illusion of determinism to the clear vision of Urania. Confined within the universe man is neurotic, for its stasis repels him: the paradisal garden of Granusion is happiest, says Bernardus, in that it "knows not that its happiness is the gift of God," and as a rather cryptic reference reminds us, man's sojourn there was inevitably brief.[181] "Imbecile" in the inadequacy of his faculties to order his life, yet capable of knowledge and aspiration, he seems fated to strive perpetually toward a fulfillment which lies beyond the limits of Nature. The tension and the futility of such a life are brought to a head in the final paragraphs of the *Cosmographia,* where the static self-sufficiency of Nature is contrasted with the exhausting labor of man:

The nature of the universe outlives itself, for it flows back into itself, and so survives and is nourished by its very flowing away. For whatever is lost only merges again with the sum of things, and that it may die perpetually, never dies wholly. But man, ever liable to affliction by forces far less harmonious, passes wholly out of existence with the failure of his body. Unable to sustain himself, and wanting nourishment from without, he exhausts his life, and a day reduces him to nothing.

The view of life thus defined is rigorous and challenging. Between the experience of fallen man, virtually imprisoned by the natural order, and original man, lord of this order and capable of comprehending it in its divine totality, there are, it would seem, only the tenuous and uncertain links represented imaginatively by the genius figures of the poem. But there is a point of view from which the very existence of these genii, these links between physical and metaphysical reality, between reason on the one hand and the irrational and supra-rational on the other, can be seen to intimate a more comprehensive and unified conception of life than this contrast between cosmic and human existence, taken by itself, implies. The essence of life, as suggested by Bernardus' treatment of Silva, is not order but aspiration, and if we consider the role of the genii not simply as serving to point up ironically the discontinuity in human life, but as reflecting an impulse toward unity, we may see a new dimension of meaning in the poem. A key to this view is provided, I think, by the system of the *De divisione naturae* of Eriugena.

Founded on the pseudo-Dionysius and the Greek fathers, and emphasizing their neo-Platonic aspects, Eriugena's philosophy considers all life as emanating from primordial causes existent in the mind of God, and as imbued with a "vitalis motus" or "actio" through which it seeks to reunite with its source. This "actio" extends through all the levels of creation, so that matter itself, as has been shown above, is pervaded by the will to ascend the scale of being. Matter itself it founded on the "coitus" of primordial causes which in themselves are "intelligibilia," and thus its essential nature is superior to all accidents of disposition.[182] Similarly man in his essential being is a trinity of faculties, on which our present gross material body has been imposed as a punishment.[183] But this original trinity exists intact and inviolable within us; our "creata sapientia" knows all things in their eternal reality as does the divine wisdom, although, like God's wisdom, it manifests itself to our conscious minds only "quibusdam signis." [184] We may well recall the "igniculi" which, in Eriugena's gloss on Vulcan's gift to Psyche in the *De nuptiis,* preserve a link between mortal man and his original dignity. This image is surely in the philosopher's mind when he describes how man's failure to honor that purer body with which God had originally endowed him led to the humiliation reflected in his present coarse body, and he was "exhorted to recover the pristine dignity of his nature through self-recognition and self-abasement." [185] Eriugena goes on to speak of the "deadly sweetness of earthly delights," which shrouds the mind and blinds it to the truth, his language echoing Martianus' description of the "Mortifera delectatio" instilled in Psyche by Venus. It is both possible and natural for man to adhere to God, though in his mortal condition he does not do so,[186] and although irrationality and "adverse virtues" are continually resisting his aspirations, he preserves the basic "actio" of all life toward God.[187]

This is not, of course, an adequate summary of Eriugena's system, but may serve to suggest the relevance of that system to the *Cosmographia*. Most obviously, as I have suggested already, it accounts for the dynamic aspect of Bernardus' Silva; we may also see in the affinities of the sun, Endelechia, and Urania, a reflection of their participation in a common vitality, the "motus" or "calor" which manifests itself in generation, in the perpetual subsistence of the cosmos, in the "ingenium" and aspiration of all animate life. In relation to the transcendent source and final goal of this all-embracing power, all formal existence is indeed only a stage, a manifestation of ideas which emanate from higher "causae" and will ultimately be reabsorbed with these causes in the divine unity. As with the mystical conceptions of the pseudo-Dionysius, the historical dimension of Eriugena's system, and the role of Christ in the restoration of all life to the divine unity, are difficult to trace, but for the purposes of Bernardus, the poet of creation, this is less important than the fact that this emanationist conception gives continuity and purpose to the striving of man's earthly nature, the struggle of his psycho-physical "genius" against the adversities of necessity. It confirms as well the reality of the transcendent goal toward which he labors, and vindicates on the most fundamental level the impulse which seeks to realize this goal. The two poet-philosophers, in short, concur in their conception of the dignity of man.

With Eriugena's system as a framework we can see the coherence

of Bernardus' universe, as expressed in the relations of his cosmic personae, in a new light. Noys and Endelechia clearly stand for more
than their strictly cosmic roles express; in the system of Bernardus'
universe they perform the functions of Plato's exemplar and world
soul, the resources of the Demiurge; but their relationship has at the
same time emanationist features suggesting rather the pseudo-Dionysian than the Platonic universe.[188] A passage from the third book of
the *De divisione naturae* provides what seems to me a valuable gloss
on their roles.[189]

When [God] first descends from the superessentiality of His own nature,
in which He may be said not to "be," He is created by Himself in the primordial causes, and becomes the principle of all essence, all life, all intelligence and all those things which "gnostic theory" considers in their primordial causes. Next, descending from these primordial causes, which
occupy a sort of intermediary position between God and creation, . . . He
is created in the effects of these and appears, manifest in His theophanies.
Hence He proceeds through the manifest forms of these effects to the lowest level of all natural life, where bodily existence resides, and thus, issuing
forth through order into all things, He creates all things and becomes all
things, and returns into Himself, recalling all things into Himself, and while
He is created in all things, yet He does not cease to be above all
things. . . .

It is possible to discern here, I think, the twofold role of Bernardus'
goddesses. Noys, in effect, is God conceived as the exemplary cause
of created life, and, in this sense, "created" by Himself. It is in Noys
that the primordial causes of all life exist, and it is Noys who is the
source both of their formal continuity as a universe, the domain of Nature, and of that allpervading "motus," that Endelechia, which suf-
fuses every corner of this universe and imbues it with an "actio" which
tends toward God. Thus all of these powers have a double aspect,
first as the principles of the Platonic cosmos, but also on a deeper
level, to which the first may be said to bear the relation of *integumentum*, as the vehicle whereby, through this cosmos, God manifests
himself in "theophany." This is the dimension of cosmic life that is
hinted at in Noys' tantalizing self definition as God's "second self,"
and in Bernardus' emphasis on the elusive character of Endelechia. It
is the essence of the vision of Urania, confirming the participation
and inevitable resolution of human life in God. Its sacramental char-

acter, and the most fundamental of the innumerable analogies which can be drawn between cosmic and sacramental realities, are conveyed at the opening of the *Cosmographia,* where Noys addresses Nature, in words which deliberately echo the angelic salutation to Mary, as "the blessed fruitfulness of my womb," thus characterizing the goddess as a type of the mother of God, and hinting at the sense in which every product of form and matter constitutes, on a level deeper than its mere formal integrity can express, an "incarnation." [190]

But as we have seen, Bernardus does not end his work with a triumphant prophecy of the ultimate resolution of all things in God. Like Milton, he concludes by stressing the challenge and hardship of life rather than its ultimate rewards. The essential punctuation for the emanationist theme, as for so many other aspects of the *Cosmographia,* is provided by the genii. It is they who provide the essential link between man's nature and the illuminated consciousness of the angelic hierarchy. They confirm as well the vestigial survival of his original comprehension of all life, and so serve to define both the potentiality and the severe limitations of man's capacity to experience that theophany which, for Bernadus, as for Eriugena and for Augustine, constitutes the essence of self-discovery.

## 8.
## The Importance of the Cosmographia

The *Cosmographia* may be said to summarize a philosophical tradition and to inaugurate a literary tradition. Bernardus' poetic allegory synthesizes more completely than philosophy itself the implications of the study of Nature as pursued in the schools of the early twelfth century, and at the same time provides a searching critique of this pursuit, its limited resources, and its failure adequately to reconcile physics with metaphysics, divine immanence with divine transcendence.[191] As I have suggested above, the Chartrian Platonic synthesis was disintegrating in Bernardus' day, and the subtle obliquity and calculated obscurantism with which he insinuates a neo-Platonist character into his ostensibly Platonic cosmology, together with his glorification of

Physis and her work, constitute a remarkably accurate index to the lines along which theology on the one hand and natural science on the other were to develop in the later twelfth century. The ideal of a harmony between macrocosm and microcosm was becoming a cliché: science was demanding practical demonstration of elemental relationships and an increasingly technical approach to astronomy and astrology; theology was setting itself to define clearly the relationship between cosmic and sacramental reality, between that integrity which reflects the orderly activity of Nature and that which is uniquely the stamp of the Logos. Science and theology become as Physis and Urania: the one is empirical, analytical, and wholly reliant on human resources for its insights into Nature. It assumes man's power to collaborate with Nature, even compete with her, and exploit her secrets in the interests of human life. In this spirit, the spirit of practical astrology, medicine, and technology, Physis' final task in the *Cosmographia* is to endow man with "all-capable hands." Theology on the other hand is concerned to reassert the attitudes of traditional Christianity in relation to human art and intuition. While exploiting the resources of science and poetry in the service of an intense concern with the ways in which sacramental reality embodies itself in the structures and events of natural life, it is concerned to emphasize in a new and radical way the breach between philosophical and religious knowledge, between the vision accessible to fallen man and that "theophanic" vision with which he was originally endowed.

The thought and poetry of Alain de Lille, an author deeply influenced by the *Cosmographia*, provides a good indication of the perspective of later twelfth-century religious thought on the philosophical and poetic world of Bernardus. The cosmology of Alain's *De planctu naturae* is charged with sacramental analogies. The domain of his Nature is a church and Genius is its priest, administering the mystical union of the Word with the *materiale verbum* of creation.[192] Bernardus' Nature is hailed by Noys in language reminiscent of the angelic salutation: on this hint Alain elaborates a conception of creation as Annunciation, the "union of Nature and the Son," the impress on Hyle of the generative kiss of Noys through the "intermediary ikon." [193]

But Alain never allows us to lose sight of the fact that these are *only* analogies, that they fall short of the meaning and efficacy of the true embodiments of sacramental reality. Alain's Genius, who stands, in effect, for all the intermediary powers which relate Nature with the wisdom of God, is explicitly contrasted with the sacramental vessel of the Word in a treatise on the angelic hierarchy. Genius is a "substantific" principle, the completion and "seal" of the integrity of the natural creature, a conception which reveals a shrewd insight into Bernardus' use of the genius figure. But this substantific integrity is capable of declaring God only in a half-complete way, and is precisely contrasted with "theophany," which reveals "the effects of the supreme cause," or in Eriugena's terms, God as he is "created" in universal life.[194]

But the *Cosmographia* did more than mark a stage in the reduction of the Arts to theology. As I have suggested, Bernardus' career seems primarily to have been passed in the schools of grammar and rhetoric. Here his influence appears in the work of the rhetoricians, Matthew of Vendôme and Geoffrey de Vinsauf, who sought to define the art of poetry for the would-be practitioner. Though concerned mainly with the technical resources of poetry, they were also provoked by a theoretical notion of aesthetic unity as a synthesis of form, content, and embellishment closely analogous to that of organic life; as, in short, a union of *integumentum* and underlying *sensus*. The *Ars versificatoria* of Matthew reveals in many points of detail the influence of the poems of Bernardus and Alain, and it is probable that the desire to formulate as nearly as possible the new richness of imagery and diction which they had introduced into the poetry of the schools was one of the main purposes of his work. Geoffrey de Vinsauf describes the poet, in terms borrowed from Bernardus and Alain, as taming unruly words, reducing them to order so that their outer *ornatus* corresponds to an ideal model.[195] Each in his way reveals a sense of the self-consistency and universality of poetry and poetic language which is due largely to the example of Bernardus.[196]

The *Cosmographia* also provides valuable evidence of the responsiveness of the schools to the emerging "secular" culture of the twelfth century, and it is possible to discern its influence in the delineation of chivalric values in the great romances of the period. Bernardus gives an important place to worldly achievement in his cosmic scheme.

Kingship and military prowess, as well as study and contemplation, find a place in the meditations of Noys and Urania's Mirror. And in the panorama of civilization described in the stars, practical and intellectual achievement are balanced, like chivalry and "clergie" in twelfth-century social thought.[197] Bernardus' ideal man is a man of both action and vision, a king as well as a philosopher, and in this respect the *Cosmographia* anticipates the *Policraticus* of John of Salisbury, and the rich Anglo-Norman culture of the reign of Henry II, pupil of Adelhard of Bath and Guillaume de Conches, and patron of the Latin and vernacular authors of his day.[198] The need to reconcile practical conduct with a philosophical conception of order is at the heart of *courtoisie* as well as Chartrian Platonism, and there is a serious nucleus of meaning in Andreas Capellanus' description of love as the power "by which the whole universe is ruled," and without which "no one in the world accomplishes anything of value." [199] *Courtoisie,* too, has its quasi-sapiential nuances, its inspirational effects are often subliminal, and it aspires always to a unified consciousness like that posited in the *Cosmographia.* To reconcile the impulse to heroic action with a sustained and stable awareness of the object of such action is the great objective of the heroes of Chrétien de Troyes, and the courtly world in which they seek to realize this goal is in effect a cosmos. Its order and hierarchy are a source of inner stability for the chivalric hero, and to stray from the standard it provides or to seek to rise above this standard is to risk the perils of the dark forest, the Silva where man's vulnerability to passion and confusion is exposed and menaced with death.

On the grounds of such affinities it has recently been argued that the *Cosmographia* was in fact a major inspiration of the large view of earthly experience presented in Chretien's romances.[200] The association, moreover, can be carried beyond considerations of motif and structure. For the range of conscious and unconscious motivation brought into implicit relation by the genii of the *Cosmographia* and the complex process of descent and reascent through which spirituality is incarnated and gradually articulated are strongly suggestive, finally, of the gamut of experience undergone by the heroes of Chretien, culminating in the experience of Perceval, who emerges from the *gaste forest* of his origins and moves steadily toward the goal of sacramental vision.

Like the clarity amid complexity of the allegories of Alain, the very unity of Chretien's romance world, the sureness with which he develops the intricate psychological dramas of his heroes and the precise movement of his narratives may perhaps be attributed in part to the soundness of the basic structure provided by Bernardus' allegory. And it helps to point up what is, perhaps, historically speaking, Bernardus' most important achievement as a poet. As I have suggested, the *Cosmographia* represents the first attempt by a medieval poet, a *modernus*, to assess and extend the classical tradition on a really ambitious scale. This is something different from the often profound insights of hagiographers into the heroic history and legend to which they opposed their stories of sainthood, and it is different as well from the brilliant "classical" poems of Hildebert and his contemporaries. The *Cosmographia* reflects nothing less than the attempt to create a new poetic world, taking the Platonic cosmology with its neo-Platonic accretions as a model, but at the same time keeping all of this continually in perspective, using it as a foil to the presentation of a larger view of reality. It is on the basis of this enterprise, this imposition of new and lofty responsibilities upon poetry, that Bernardus merits comparison with Dante. The same sense of a new poetic vision leads Alain de Lille to claim that his *Anticlaudianus*, dealing with the regeneration of man, is more truly an epic than the legend of Troy or the life of Alexander, sung by his contemporaries Joseph of Exeter and Walter of Chatillon. And the work of Alain and Bernardus, the intuition and definition of an archetypal pattern in the works of the great *auctores* and the use of this pattern as the basis for new departures anticipates in profound ways the structure and themes of the *Divine Comedy*, and the use of the figure of Vergil as a foil to that of Beatrice.

The poetic world created by Bernardus was to provide the framework for an increasing range of literary exercises as the vernacular literatures of the Middle Ages emerged. The limitations of the Chartrian Nature, and of the Platonic universe as an environment for man, are transcended in the spiritual epics of Alain and Dante; in the

*Roman de la Rose* as elaborated by Jean de Meun these limitations are exposed in another way. A meticulous refusal to "mystify" by analogy such figures as Nature and Raison, the *Roman*'s counterpart to Urania, is accompanied by an ingenious exploitation of the subliminal appeal of the figure of Genius, whose relations with both the paradisal origins of man and his sexual nature are magnified and juxtaposed in a tour de force of brilliant comedy. The effect is virtually to deny the stabilizing and enlightening function of the Chartrian Nature, to suggest, on the one hand, that her order and continuity are achieved irrespective of any conscious participation on the part of man, and at the same time to emphasize in a new and powerful way the need for aid from some higher power, if man is to realize his true nature and destiny.

In a sense, Jean's vigorously pragmatic approach to Bernardus' goddess and her domain is a development of the implications of Physis and her work; his Nature attempts to deal with her problems through experimental science, and her relations with Genius, and through Genius with man, are practical in the extreme. Her vitality and activity, as conveyed in Jean's dramatization of her "personality," have an almost compulsive quality. Alain's Nature, following out the implications of the Fall for the integrity of her domain, had gone beyond Bernardus' goddess in realizing the necessity of a renewal, and so had dwelt at length and in reverent humility on the mystery of "a nativity whose nature she could not understand;" but Jean's goddess, led to a similar confrontation, makes only the briefest acknowledgement of the Incarnation before becoming reabsorbed in her own concerns.[201] At this point, I think, we must recognize the influence of the far more radically scientific spirit of later thirteenth-century "natural philosophy," and a realism which appears also in Jean's treatment of human society, as having interposed themselves between his view of the world and that of the twelfth-century Platonists.[202]

I would like to conclude by considering the question of the extraordinary appeal which the *Cosmographia* has always held for its readers, few though these have been in recent centuries. Its secret, I think, lies in Bernardus' extraordinary sense of the general, his ability to see the potentially universal implications of his sources and materials, and to imbue his own work with these same qualities. It appears in his sure sense of what is archetypal and psychologically telling in models so diverse as Vergil, Ovid, Claudian, and the Latin neo-Platonists, and his ability to synthesize their contributions in a single allegorical form. It appears even in the grace with which he hints at his own "civilizing" role as the poet of northern Europe, and with no breach of decorum weaves into a survey of ancient lands and civilizations the Alps and the Pyrenees, the strange-sounding names of the forests, fauna, and flora of France, the *res gestae* of saints and the dynasties of Frankish kings.

But the most important manifestation of this instinct for the general, and also, no doubt, the quality which has led so many readers to see in Bernardus the exponent of ideals, hopes, and fears of which he knew nothing, is the psychological truth of his allegory of the creation. There is a sense of urgency and anxiety in his presentation which seems to exist, as it were, independent of the coherent "message" of the poem. In one sense this is a historical accident. The Platonist synthesis with which Bernardus worked had already, by his day, outlived its true value as a means of coordinating the wisdom of the *auctores;* it gave eloquent expression to the ideal of human self-realization without providing the means to meet its challenge; as Southern observes, it "could point no way forward." [203] The scientific revival of the late twelfth and early thirteenth centuries, which saw the development in the universities of a far more detailed and coherent view of the natural order than any accessible to the Chartrians, was only beginning. Bernardus' intensity is thus to some extent simply a reaction to the uncertainty which any period of major transition produces. But on a deeper level it is an expression of the yearning for coherence and an ideal of perfection which is implicit in any serious treatment of the idea of world harmony.[204] Disciplined and set in relation to Bernardus' broader spiritual perspective on his theme, it lends a peculiar note of irony—a stronger, more Vergilian quality than the "Christian melancholy" which Leo Spitzer detected in the poetry of Petrarch—to his exposure of the illusion of continuity and permanence in human affairs.

But if Bernardus' sense of loss is inescapable, he is equally humane and equally convincing in his insistence that life is ongoing. True to his role as the poet of creation, he is capable of conveying the intuition of an underlying spiritual *motus* without abandoning the task of expressing the perpetual struggle of cosmic life. Spirit and conflict are the two great themes of the *Cosmographia,* and it is a confirmation of Bernardus' humanism that the realization of the one demands a full and sympathetic articulation of the other.

# *A Note on the Text*

The *Cosmographia* survives in more than fifty manuscripts, which exhibit certain variations in form. Some give only the verse portions of the work, and several offer variant versions of parts of the catalogue poem 1.3. The text of the difficult philosophical poem 2.8 became corrupted at an early stage, and some twelfth-century manuscripts omit it altogether. There is, though, no reason to doubt that the version printed by Barach and Wrobel, which is the fullest, is also the most authentic.

However, the Barach-Wrobel text was carelessly edited from inferior manuscripts and does not, by itself, provide an adequate basis for a translation. I have therefore drawn on several French and English manuscripts of the work, and chiefly Oxford Bodleian Laud Misc. 515. I have also used a mimeographed text of the *Megacosmos* prepared by Mr. Peter Dronke, who collated the Barach-Wrobel text with that of the Laud manuscript. I have examined the fine critical edition of the *Cosmographia* prepared by Professor André Vernet (diss. Ecole des Chartes, Paris, 1937), but have made only the most sparing use of it, preferring not to anticipate its appearance in printed form.

I append a list of the most important new readings, nearly all of them based on the Laud manuscript, that I have introduced in adapting the Barach-Wrobel text for purposes of translation.[1] I have not bothered to note my many disagreements with its extremely erratic punctuation.[2] Numbers refer to page and line in Barach-Wrobel.

7.9 imagine *for* imaginem
8.38 quae *for* et
8.39 intra *for* infra
8.55 cum sis *for* causis
8.59 recedat *for* recedit
9.3 evocata *for* evocato
10.30 parte pro plurima *for* per te procul pellam
10.42 dedignor *for* dediger
11.85 junxit *for* vinxit
13.166 contiguum *for* congruum
13.168 iubarque *for* iubar et
15.210 per naturam *for* personam
18.120 Exaequans *for* exequias
19.147 plenius *for* plenus
21.218 angusta *for* angusto
22.232 After this line, the following couplet: Carior et redolens et burse predo sabellus / Gutturaque amplectens deliciosa ducum.
22.259 germina *for* nomina
24.300 quaecumque *for* quamcumque
27.411 iusquiamo *for* ius cyamo
28.438 After this line, the following couplet: Commendat Ligerim darsus parilisque saporis / Stagna lacusque timens longior umbra natat.
28.437 Tructa *for* turtur
29.477 quem *for* quam
31.61 Rerum *for* Verum
32.98 resolvitur *for* revolvitur
32.109 est *for* et
35.4 artificis *for* artifices

39.28 injuncti *for* invicti
42.67–68 impostum *for* in posterum
43.116 quarum *for* quarta
45.192 intervius *for* interius
47.1 Homeri *for* homini
48.39 quid *for* quod
49.53 octava *for* orta
51.4 contiguamque *for* contingantque
51.13 quae Mercurii *for* Mercuriique
51.16 fati *for* facies, probas *for* probat
51.17 constringat *for* confringat, ver *for* ut
51.19 mollisque *for* mollisve
51.21 novetur *for* movetur·
51.22 partu turgeat omnis humus *for* partum urgeat omnis humum
51.24 ut *for* et
51.32 Divisumque genus dividit illud idem
51.34 quasi *for* quia
52.37 actor *for* amor
52.41 inspirat verum *for* inspiciat quatuor, conscia for consequa
52.49 falso *for* flore
52.50 pervideas *for* provideas
53.32 gremio *for* genio
60.37 Aevi *for* Cui
61.55 supernos *for* superbos
61.58 agit *for* aget
62.50 materiali *for* mali
66.30 tereti *for* tenti
67.61 quasi *for* quia
70.164 Non a *for* nam

## The Form of the Cosmographia

Like Martianus' *De nuptiis,* Boethius' *De Consolatione,* and the *De eodem et diuerso* of Adelhard, the two books of the *Cosmographia* are composed of alternating verse and prose passages of irregular length. Nature's appeal in 1.1 is in dactyllic hexameter. The catalogue poem 1.3 and sections 2.2, 2.6, 2.8, 2.10, and 2.14 are in elegaic couplets. Section 2.4 consists of alternating hexameter and tetrameter lines, apparently in imitation of *De consolatione* 1. metr. 3, and 2.12 is in the "Archilocheian" meter of Horace, *Odes* 4.7. The remaining sections are in prose.

BERNARDUS SILVESTRIS

# Cosmographia

# Dedication to Thierry of Chartres

To Thierry, doctor most renowned for true eminence in learning,[1] Bernardus Silvestris offers his work.

For some time, I confess, I have been debating with my innermost self, whether to submit my little work for a friendly hearing or destroy it utterly without waiting for judgment. For since a treatise on the totality of the universe is difficult by its very nature, and this the composition of a dull wit as well, it fears to be heard and perused by a perceptive judge.[2]

To be sure, your kindly willingness to inspect a piece of writing lacking in art, but dedicated to you, has aroused my boldness, quickened my spirits, and strengthened my confidence. Yet I have decided that a work so imperfect should not declare the name of its author until such time as it shall have received from your judgment the verdict of publication or suppression. Your discernment, then, will decide whether it ought to appear openly and come into the hands of all. If meanwhile it is presented for your consideration, it is submitted for judgment and correction, not for approval.

May your life be long and flourishing.

# Summary

In the first book of this work, which is called *Megacosmos*, or "the Greater Universe," Nature, as if in tears, makes complaint to Noys, or Divine Providence, about the confused state of the primal matter, or Hyle, and pleads that the universe be more beautifully wrought. Noys, moved by her prayers, assents willingly to her appeal, and straightway separates the four elements from one another. She sets the nine hierarchies of angels in the heavens; fixes the stars in the firmament; arranges the signs of the Zodiac and sets the seven planetary orbs in motion beneath them; sets the four cardinal winds in mutual opposition. Then follow the creation of living creatures and an account of the position of earth at the center of things. Then famous mountains are described, followed by the characteristics of animal life. Next are the famous rivers, followed by the characteristics of trees. Then the varieties of scents and spices are described. Next the kinds of vegetables, the characteristics of grains, and then the powers of herbs. Then the kinds of swimming creatures, followed by the race of birds. Then the source of life in animate creatures is discussed. Thus in the first book is described the ordered disposition of the elements.

In the second book, which is called *Microcosmos,* or "the Lesser Universe," Noys speaks to Nature, glories in the refinement of the universe, and promises to create man as the completion of her work. Accordingly she orders Nature to search carefully for Urania, who is queen of the stars, and Physis, who is deeply versed in the nature of earthly life. Nature obeys her instructress at once, and after searching for Urania through all the celestial spheres, finds her at last, gazing in wonder at the stars. Since the cause of Nature's journey is already known to her, Urania promises to join her, in her task and in her journey. Then the two set out, and after having passed through the circles of the planets and forewarned themselves of their several influences, they at last discover Physis, dwelling in the very bosom of the flourishing earth amid the odors of spices, attended by her two daughters, Theory and Practice. They explain why they have come. Suddenly Noys is present there, and having made her will known to them she assigns to the three powers three kinds of speculative knowledge, and urges them to the creation of man. Physis then forms man out of the remainder of the four elements and, beginning with the head and working limb by limb, completes her work with the feet.

# Megacosmos

**Chapter One**

When Silva, still a formless chaotic mass, held the first beginnings of things in their ancient state of confusion,[1] Nature appeared, complaining to God,[2] and accusing Noys, the unfathomable mind: "O Noys, supreme image of unfailing life, God born of God, substance of truth, issue of eternal deliberation, my true Minerva.[3] Though what I seek to realize be beyond my comprehension—that Silva be made more malleable, that she cast off her lethargy and be drawn forth to assume the image of a nobler form[4]—yet if you do not consent to this undertaking I must abandon my conceptions.[5]

"Surely God, whose own nature is supremely benevolent, generous, and not liable to the miserable agitation of envy, wills the melioration of all things, so far as their materiality will allow;[6] the author does not disparage his work.[7] Thus you cannot be envious, though you should bestow upon the unwieldy mass a full and perfect grace, if I recall truly the hidden ways of your deliberation.

"Silva, intractable, a formless chaos, a hostile coalescence, the motley appearance of being,[8] a mass discordant with itself, longs in her turbulence for a tempering power; in her crudity for form; in her rankness for cultivation.[9] Yearning to emerge from her ancient confusion, she demands the shaping influence of number and the bonds of harmony.[10]

"Why then do inborn conflicts and wars among kindred properties assail foundations established from eternity by the first cause? [11] When the mass ebbs and flows, at odds with itself, when hapless elements are borne about at random and the whole body is rent by sudden agitations, what does it avail Silva, mother of all, that her birth preceded all creation, if she is deprived of light, abounds only in darkness, cut off from fulfillment?—if, finally, in this wretched condition, her countenance is such as to frighten her very creator?

"Hyle presents herself at your feet with all her progeny, to express to you her resentment that, though grown whiteheaded, she has lived out her age in formless squalor. Behold, oldest of things as she is, Silva yearns to be born again, and to be defined in her new birth by the shapes of creation.

"No small honor and thanks are due to Silva, who contains the original natures of things diffused through her vast womb. Within this cradle the infant universe squalls,[12] and cries to be clothed with a finer appearance. The tender world gives vent to these tears that it may come forth from the bosom of its nurse and forsake fostering Silva.

"The elements come before you, demanding forms, qualities, and functions appropriate to their causal roles,[13] and seek those stations to which they are almost spontaneously borne,[14] drawn by a common sympathy: lively fire to the height, heavy earth downward, moisture and air abroad through the middle region.

"Why then does this chaotic mass draw all things together in confusion? It is Silva's plight to be whirled about in flux and thrown back again into her original confusion by random eddies; peace, love, law, and order are unknown to her. Because she is lacking in all these Silva may scarcely lay claim to her true title as the work of God; rather she appears a giddy contrivance of blind fortune, bereft of the protection of any higher power.

"By your leave, bountiful Noys, let me speak: supremely beautiful though you are, your rule over Silva, your dominion, is exercised in a barren and ugly court; you yourself seem old and sad. Why has privation been the companion of Silva from eternity? Let it depart, through the imposition of order and form.[15] Apply your hand, divide the mass, refine its elements and set them in their stations; for they will appear more pleasing when thus disposed. Quicken what is inert, govern what moves at random, impose shape and bestow splendor. Let the work confess the author who has made it.

"It is for the universe that I, Nature, appeal; it is enough for me if I may behold the birth of the universe and its creatures, I seek no more. But what more can I say to you? I blush to have sought to instruct Minerva."

*Chapter Two*

Here Nature made an end. Then Noys, her countenance brightening,[16] raised her eyes to the speaker, and, as if summoned forth to discourse from the inner chambers of her mind,[17] replied: "Truly, O Nature, blessed fruitfulness of my womb,[18] you have neither dishonored nor fallen away from your high origin; daughter of Providence as you are, you do not fail to provide for the universe and its creatures.[19]

"And I am Noys, the consummate and profound reason of God,[20] whom his prime substance brought forth of itself, a second self, not in time, but out of that eternal state in which it abides unmoved.[21] I, Noys, am the knowledge and judgment of the divine will in the disposition of things. I conduct the operations over which I preside accordingly as I am bidden by the harmonious action of that will.[22] For unless the will of God be sought, and until his judgment concerning created existence is pronounced, your haste to bring created life into substantial existence is vain. The nativity of creatures is celebrated first in the divine mind; the effect which ensues is secondary. Thus the plan of a cosmogony, which you had conceived by sacred and blessed instincts and in accordance with a higher plan, could not be brought to present realization until the term established by higher laws. Unbending and invincible necessity and indissoluble bonds had been imposed, so that the cultivation and adornment which you desire for the universe might take place no earlier. Now at last, because you appeal at the proper time, and appeal in behalf of causes which concur in the impulse to order, your desires are served.

"Now Hyle exists in an ambiguous state, suspended between good and evil, but because her evil tendency preponderates, she is more readily inclined to acquiesce in its impulses.[23] I recognize that this wild and perverse quality cannot be perfectly refined away or transformed; being present in such abundance, and sustained by the na-

tive properties of the matter in which it has established itself, it will
not readily give way. However, so that the evil of Silva may not dis-
rupt my work or the order I impose, I will refine away the greater
part of her coarseness. Moreover, the teeming mass, now violently assailed by a restless movement born of confusion, will be reduced to
carefully established confines by that peace which is my special concern. I will produce a form for Silva, through union with which she
may come to flower,[24] and no longer cause displeasure by her ill-or-
dered appearance. I have ordained that her substance be refashioned
in a better condition. I will instill amity in the universe and regularity in the elements. I am indeed troubled that the initial state of creation should be so deprived. But the emergence of shape will remove
this privation from material existence.

"Accordingly I will now begin the work which I have promised, for
he who acts too slowly torments those who await the issue. And since
you, Nature, are well endowed with intelligence, and seek to realize
this object through your prayers, I will not scorn to accept you as ally
and companion in the task."

Nature stood alert, her mind intent upon the voice; for this speech
wrought out of things she had hoped for was delightful to her ears.
And when she understood that what she had desired was granted, she
bowed low before Providence, grateful in mind and countenance
alike, and threw herself at her feet.

Hyle was Nature's most ancient manifestation,[25] the inexhaustible
womb of generation, the primary basis of formal existence, the matter
of all bodies, the foundation of substance.[26] Her capaciousness, confined by no boundaries or limitations, extended itself from the beginning to such vast recesses and such scope for growth as the totality of
creatures would demand.[27] And since diverse and intricate qualities
pervaded her, the matter and foundation of their perpetuity, she
could not but be thrown into confusion, for she was assailed in such
manifold ways by all natural existence.[28] The numerous and uninter-
rupted concourse of natures dispelled stability and peaceful repose,
and departing multitudes only afforded space for more to enter. Hyle
existed without rest and could not remember a time when she might
have been less continually engaged in the formation of new creatures
or the reassimilation of those deceased. Vacillating, and ever liable to

change from one state of quality and form to another, no material nature might hope to be assigned an identity proper to itself, and so
each went forth unnamed, putting on a borrowed appearance. And
since it was liable to assume any shape, it was not specially stamped
with the seal of a single form of its own.[29]

Yet this freedom to move at random was restrained by a certain
agreement, in that the restless material was sustained by the more
stable substantiality of the elements, and clung, as it were, to these
four roots.[30] Because of this Silva might more safely suffer herself to
be drawn out and enlarged by an infinite range of essences, qualities,
and quantities. Yet to the very extent that her nature proved fertile
and prolific in conceiving and giving birth to all creatures in common, it was equally impartial with respect to evil.[31] For there was in-
fused in her seed-bed from of old the taint of a certain malign tendency which would not readily abandon the basis of its existence.
The seeds of things, too, warring with one another in the chaotic
mass—fiery particles with icy, the sluggish with the volatile—dissipated the material or substantial qualities of their common subject matter by the clash of their contradictory tendencies.

Accordingly divine Providence, to remedy this condition by the
promised transformation, reviewed the resources of her mind, mustered her faculties, and summoned up her imaginative powers. Since
the reconciliation of discords, the aggregation of incongruities, and
the yoking of mutually repellent forces seemed to be the only principles of arrangement, she resolved to separate mixed natures, to give
order to their confusion and to refine their unformed condition.[32] She
imposed law and restrained their freedom of motion. Rude though
they were, she effected a balance of properties among her undisciplined and recalcitrant materials, joined them with means, and so
bound them together in arithmetical proportion.[33] As the bonds of a
reconciling concord, sprung from the inner deliberations of Providence, were thus interposed, the rough and, as it were, the uncivil-
ized strain in matter changed its obstinacy to cooperation, and submitted its innate conflict to a general reconciliation. Once this rigidity
of ancient, even primordial lineage had been overcome, an adapt-
ability took its place capable of being drawn into such channels as
Providence decreed.

When she had so nearly refined away that coarseness which is the property of Silva, Noys, reflecting inwardly upon eternal ideas,[34] fashioned the species of created life in close and intimate resemblance to these. Hyle, who had lain shrouded in dullness and obscurity, assumed a different aspect, once given definition by visible images of the ideal.[35] When the mother of all life thus gave scope to the fullness of her generative capacities, and opened forth the womb of her fecundity to the production of life, there straightway took place, from this source and within it, the origin of the created essences, the birth of the elements.

From the confused and turbulent depths the power of fire emerged first, and instantly dissipated the primeval darkness with darting flame. Earth appeared next, distinguished by no such lightness or radiance, but stable in tendency, and of a more concrete corporeity; for she was destined to reclaim, once their earthly round was completed, the returning stream of all those creatures which would be born of her. Forth came the gleaming substance of clear water, whose level and shimmering surface gave back rival images when darkened by the intrusion of shadows. Then the vast region of the air was interposed, volatile and subject to change; now giving itself to shadows, now gleaming at the infusion of light, now growing crisp with frost, now languid with heat.

When each of these bodies had taken up the abode to which it was most readily drawn by material affinity, the earth rested firm, fire darted far above, and air and water assumed intermediate positions.[36] This balancing and mediating tendency was interposed so that under its peace-making influence the elements, by imposing boundaries on themselves, might establish friendly and cooperative dominions.[37] For example, fire, hotter and more volatile than the others, might have rejected the constraints of the established order had not air and water, allied by their kindred properties, pledged their mutual assistance to resist it. Moisture opposed itself to dryness, gravity offset the tendency to instability. The earth, when parched, relubricated its generative channels by means of the waters which impinged upon it, and sustained itself by the infusion of air, lest it sink below its ordained position, weighed down by corporeal substance. Thus there was no way for discrepancies among the diverse kinds of life to introduce a discrepancy in the total scheme, where discrepancies were reconciled.

Thus the contentious and discordant multitude of warring factions, as though laying aside their arms, entered into a condition of peaceful unity. Noys began to review the first products of her partly completed labor. She saw that all that she had made was good, and would be pleasing in the sight of God; [38] for species had emerged from the refinement of matter, a median tendency from the imposition of law, stability from materiality, a single fullness from many parts. And indeed it was inevitable that these components should form a full and consummate whole, since fire, earth, and the other materials had themselves wholly consented, in their essential being and in their properties, to a full and consummate fusion of their own.

Now if in the course of the work a single particle, however small, had been left out of this synthesis of matter, the cosmic order which was about to be realized might thereby have been subject to disruption and damage, since it would manifestly have been liable to assault from outside by foreign bodies.[39] So now the hapless race of men, whose existence is not sustained by a perfect fusion of the elements, lives in constant fear of falling subject to external accidents.[40] For whenever heat from without aggravates the heat of man's nature, his inner peace is disrupted, and what had existed in a state of calm becomes aroused to destructive activity.[41] Therefore provision was made in the divine plan that whatsoever in the temporal order might violate the scheme of the universe, disorder its substance, or interfere with its operation should be cut away with the sources of its activity and destroyed.

When these necessary steps had been taken with regard to matter, when the framework of the elements was now solidly established, the outward shape of creation made beautiful and its coherence become a very miracle, Noys turned her intelligence from Silva to the production of a cosmic soul.

She was the fountain of light,[42] seed-bed of life, a good born of the divine goodness, that fullness of knowledge which is called the "mind" of the most high.[43] This Noys, then, is the intellect of supreme and all-powerful God, a nature born of His divinity.[44] In her are the images of unfailing life, the eternal likenesses, the intelligible universe, sure knowledge of things to come.[45] There, as though in a clearer glass, might be seen all that God's hidden will would bring to pass through temporal generation, or by divine act.[46] There were enrolled, in kind, in species, in individual uniqueness, all that Hyle, that the cosmic order, that the elements labor to bring forth. There, inscribed by the finger of the supreme arbiter, were the fabric of time, the chain of destiny, the disposition of the ages.[47] There were the tears of the poor and the fortunes of kings, the soldier's strength and the happier discipline of the philosophers, all that the reason of angels or men may comprehend, all that is gathered together beneath the dome of heaven. What exists in this way is at one with the eternal, of the nature of God and inseparable from Him in substance.

From the very source, then, of this our life and light, there issued forth by a sort of emanation [48] the life, illumination, and soul of creation, Endelechia. She was like a sphere, of vast size yet of fixed dimensions, and such as one might not perceive visually, but only by intellect. Her shining substance appeared just like a steadily flowing fountain, defying scrutiny by its uncertain condition since it seemed so closely akin to the atmosphere, and at the same time to the heaven itself. For who has defined with certainty that mode of being which emerges from harmony, from number? And so, when one was deluded as if by magic as to its true aspect, it was beyond the reach of scrutiny to divine how this vitalizing spark [49] should so endure that it might not be extinguished, but was rendered back whole and undiminished by each individual creature.

Now by her birth this Endelechia was closely and intimately related to Noys. Lest so glorious a bride should protest that the universe spawned by mother Silva was an unworthy husband, Providence arranged the terms of a special compact, wherein material and heavenly nature might arrive at a consistent harmony by way of congruent proportions. And since what is subtly refined does not willingly accord with what is dull and heavy, a more adaptable mean proportion interceded to effect their connection, and fastened body to soul as if glued, or bound in marriage. Thus when their normal hostility had been changed to liking, and each took pleasure in the other, agreement gave birth to amity, and amity to trust—as may be seen to this day.[50]

The universe labored in ceaseless pain [51] under the affliction of the poundings and vexations which it suffered whenever an irruption or inundation, due to an excess of heat or moisture, disturbed the accustomed course of nature.[52] As swiftly as possible Endelechia attacked this problem, and labored to repair her dwelling place. The rites of hospitality were maintained, but she would neither cooperate with nor suffer willingly any unbidden intruder within her tabernacle.[53]

When an alliance had thus come about by mutual agreement of soul and cosmos, the newborn universe quickly transmitted the initial spark of life, which it had received through the infusion of spirit, from the total structure to individual existences, by the mode of animation or vegetation that best suited the special capacity of each.[54] Ethereal existences received ethereal spirit, pure was mated with pure, nor was it strange that each nature should embrace most firmly a spirit consonant with it.

The true affinity of Endelechia is with the firmament and the stars, so that in the supernal realm her power to sustain celestial life endures undiminished, while at the lower levels of existence its efficacy declines. This weakening accounts for the sluggishness of bodily existence, wherein Endelechia appears less powerful than she is in her true nature.

And now, made strong by the vivifying gift of soul, the totality of created life unfolded in ordered progression[55] from the nurturing womb of Silva.

## Chapter Three

Thus the subtle ether grew separate from the stars, the stars from the heavens, the heavens from the atmosphere, the earth from the deep. In the heavens and with the aid of heavenly powers the hand of God produced the first fruits of the great work of creation: the rounded form of the celestial sphere, the purer essence of fire,[56] circular motion, the host of divine powers.

I call "gods" those beings whose presence is ever attendant upon God,[57] those whom true day preserves in its true light. For a region of calm, exempt from all the tumult of the atmosphere, secluded unto itself, sets apart their secret dwelling places. Far on high, beyond the limits of the universe, belief places the being of God.[58]

Perfect in understanding, the Cherub beholds most directly and fully the hidden deliberations of God. Far different is the burning desire of the Seraphim, but God is the desire of these subjects too, and their desire is a sacred love. The Thrones are that pure host among whom resides the unfathomable spirit and understanding, the profound mind of Noys.[59] That throng mighty in command by right of power bear a name which is the emblem of their office. To those spirits whose special dominion provides their name is subordinated an order inferior in condition; but although subordinate to those above, the Prince assigns duties to subordinate ranks, and issues commands of his own. The sacred order of Virtues bring miracles to pass, for they possess the principles of their own function.[60] A portion of the heavenly army as numerous as the stars are those who serve obediently at the bidding of Michael. The lowest grade is that of the Angels. Add those above: [61] their ranks divide into nine hierarchies.

Far removed from earthly existence, the substance of the heavens, finer in composition, was given a finer ornamentation. For the firmament is inscribed with stars, and prefigures all that may come to pass through decree of fate. It foretells through signs by what means and to what end the movement of the stars determines the course of the ages. For that sequence of events which ages to come and the measured course of time will wholly unfold has a prior existence in the stars. There are the scepter of Phoroneus, the conflict of the brothers at Thebes, the flames of Phaethon, Deucalion's flood. In the stars are the poverty of Codrus, the wealth of Croesus, Paris' incontinence, the chastity of Hippolytus, Priam's pomp, the boldness of Turnus, Odyssean cleverness, and Herculean strength. In the stars are the boxer Pollux, Tiphys the helmsman, Cicero the orator, and the geometrician Thales. In the stars Vergil composes with grace, Myro creates forms, Nero shines in Latian nobility. Persia charts the heavenly bodies, Egypt gives birth to the arts, learned Greece reads, Rome wages war. Plato intuits the principles of existence, Archilles fights, and the bountiful hand of Titus pours out riches. A tender virgin gives birth to Christ, at once the idea and the embodiment of God, and earthly existence realizes true divinity.[62] Divine munificence bestows Eugene upon the world, and grants all things at once in this sole gift.[63] Thus the Creator wrought, that ages to come might be beheld in advance, signified by starry ciphers.

Noys set firm the twin poles, and caused the revolving firmament to move about them; and this revolution was its eternal state.[64] With five parallel zones she encircled our middle orb. Because of these its extremes are frozen and its central portion fiery hot; and thus she created two temperate zones, between the coldness of the extremes and the sun's unswerving course across the central region. She divided the heaven into quarters with encircling colures (but neither of them attains the point of completion),[65] and set out the circle of the Zodiac; its greater arc extends southward, the lesser toward the starry team and their icy wagon. Likewise that milky band whose radiance is produced by clustering stars was cast across the sky. The point of the Solstice appeared close to Cancer, and Libra became the equidial boundary.

Draco, passing between, separates the two Bears. The northern pole is marked by the mariner's star, which may never behold its own antipodes. Bootes attempts brief sorties from the upper region of the sky. The Kneeler travels the same regions of the sky as Helice and the lesser Cynosura. Behind his Herculean shoulders gleams the crown of Ariadne, and Lyra, the discovery of Mercury, lies before. Next in position comes the Ledean swan. Cepheus and Cassiopeia follow in close succession. An effulgence of milk-white gold marks the midpoint of Andromeda; Perseus holds up the face of a glowing Gorgon. At the birth of those Kids whom the Charioteer carries, many days were filled with rain. In the next position among the stars shines Ophiucus, master of the powers of herbs, his tender body girt by an unyielding snake. The brilliant Serpent is stretched forth to the full extent of its vast body in a still more splendid pattern of stars. The Arrow, also remarkable for its outstanding splendor, burns brightly. Lower down hovers the bird of Jove with curving beak. The Dolphin impinges closely upon the domain of Aquila, and the horse of Bellerophon stands hard by the Dolphin. Set above your vessel, O Phrixean, shines that deltaic form which indicates the position of Egypt.[66] The weeping Hyades are set in the forehead of the Bull, and seven sisters form his tail—call them the Pleiades or the Vergiliae. Prochius brings on the blighting heat of summer at that time when the Zodiac reaches its height in Gemini. Sirius, notorious for heat, burns in baleful brilliance at the point of the Solstice in Cancer. The heavenly Manger reveals the two Aselli, set before Cancer and the Herculean monster. Here too Orion hunts across the starry regions, and the Hare is quick to anticipate his course. The famous Argo, which first ventured upon the deep, pursues a heavenly course with no Tiphys at the helm. The Altar, set still lower in the sky, inclines toward the southern region, close by Sagittarius, an old man half a beast.[67] The seagoing whale occupies the next position, where the Fish are joined with one another by a celestial bond. Eridanus, famous in our clime, flows also on high, and bears a name not unknown to the gods. The sprawling Hydra, and the Bowl, placed below the seat of Cancer, occupy a common position with the Crow. But the Fish takes in the waters of brimming Aquarius.[68] Other signs shine forth in their several places.

Poised in opposition to the Phrixean Ram, Libra gives forth brilliant beams, setting day and night in equal measure. Scorpio, troubling by the very depravity of his nature, rages with fierce glares against his opposite, the Bull. The keen arrow of the Thessalian Sage is drawn against the glowing stars of the twin brothers of Helen. The Goat, so devoted to Jove in his Cretan boyhood, shines with the brilliance of the shifting star of Cancer, or the ravaging Dog. Cancer rages with scorching heat, and Capricorn, at the opposite solstitial boundary, brings drenching rain. The little vessel of a boy pouring forth water counters the might of Hercules and the raging Lion. The Fish, as they rise, behold the decline of Astraea, who gathers in the wealth of the fruitful year.

The stars, which the present age calls by this name or that, existed at the birth of time as heavenly fire. Lest he should stumble in seeking to express this universal theme in common speech, man created those names which even now denote the stars.[69]

Below the heaven where the signs move about an inclined path determines the courses of the seven planets. Noys preordained their nature and behavior, and what influence each in its own right would exert on the world. The Moon passes close to the earth, and makes the sea rise up against the land by her imperious decree. He who comes next, by a system of widely varying movements, passes back and forth across the path of the Sun. Next in position is Venus, suffused with humid heat, who has charge of the vital powers of sowing and generation. The splendor of the Sun is in the central position, that the stars on every hand, sustained by the god of light, may give off a fuller radiance. Mars, following next after the Sun, visits war upon proud cities, and his red glare often works strangely upon kings. Sixth is the bounty of Jove, which, where not tainted by some extraneous evil, brings favoring signs to fulfillment in happy events. In the outermost sphere the star of old Saturn, barren and cold, moves in wide and sluggish circles.

Noys commanded the winds, born of the shifting sea, to blow in opposition to one another from established positions. The north wind is crisp with frost, the south heavy with rain, the east wind brings storms, the west fair weather.

Now when the earth stood firm beneath the heavens, the sea ebbed and flowed, and the starry ether gave off a new radiance, the beasts of the field and the reptiles, winged creatures, and fishes were made, and took over the regions proper to their diverse conditions. Whether it creep or walk, swim or fly, each lives by its own law, and none agree as to their mode of life. For savage beasts are born to the wilderness, others to the field, the serpent to the barren ground, the bird to the air, fish to the waters. The waters are swum by fish, the air is traversed by birds, beasts walk abroad, the viper creeps along the ground.

The earth had settled at the center of things, seemingly reduced to a point; rest from all motion was afforded by its stable position. Marked by three great divisions, all the land either withdrew from sight or, in scattered spots, lay open to the seven climates.[70] One part was covered by waters, another by wilderness; the small expanse of land remaining was left bare.

The earth was bound by mountains as if with sinews. Atlas sup- , ports the firmament and all the stars. At the very threshold of the ethereal region renowned Olympus beholds the dense clouds of a lowering sky. Parnassus with its twin peaks seeks to behold the gods disposing the affairs of men and the seven planets. Lebanon bristles with cedars; Sinai, where the blessed Law was given into the charge of blessed Moses, lies open. Athos rises, together with Eryx and lofty Cythera, Aracyntheus, and the peak of Aganippe; the Appenines too, and Oeta, sepulchre of Hercules; glowing Lipara, hills redolent of terebinth;[71] Pindus and Ossa, a peak menacing to the gods; Othrys, and Pelion, retreat of the ancient healer; the Caucasus, watchtower of stargazing Prometheus; Rhodope, a clime favored by him who sang to the lyre. The Italian land swells to the height of Gargano, and Sicily is exalted by your peak, O Peloris. Pholoe, known for having borne the biform race of Centaurs, threatens heaven with her double peak. Snows which Boreas brings to birth in that clime whiten the peaks of the northward-lying Rhipean mountains. Their very situation links the Alps together on every side, and toward the setting sun narrow passes lie frozen.[72]

Such open ground as the mountains enclose never knows the plow, and the land of their narrow valleys lies idle. For their thickets harbor wolves, their deserts lions, the dry wastes serpents, and the woodland boars.[73]

Genus is drawn out into species, and nature, simple and one herself, is particularized in diverse ways.

The elephant is fortified with tusks,[74] the camel's back rises high, and horns grace the forehead of the gazelle. The stag is equipped for flight, and the doe lifts her long slender legs with knees drawn high. The bold lion relies on his stout heart, the bear on his claws, the terrible tiger on his fangs, the fearsome boar on his tusks. The sheep becomes soft with fleece; the she-goat and her spouse are clad in a coarser robe. An impetuous heart drives on the horse, but sluggishness burdens the donkey; the weight of his ears lies heavy on his spirit. The panther and the bloodthirsty wolf howl for prey; the one loves the forest, the other the mountain ridge. Though the bull's is a nobler spirit, the fox has a fuller store of understanding within his small thin frame. Oxen are born to slavery, and that creature of fear, that fugitive thing, the hare, waxes mighty in ears. The wild ass, fleeing to the mountains, abandons the exercise and repudiates the obligation of his bodily nature.[75] Whether his affection be due to native understanding or to habit, the dog submits to living in terror of human threats. The lynx comes forth to create miracles which none may behold; for he possesses within himself a fountain of liquid light.[76] The ape comes forth to receive men's laughter, a deformed image, a man of degenerate nature. The beaver comes forth, prompt to give up from his own body those treasures which a greedy enemy pursues.[77] The squirrel creeps forward, and the marten, destined to clothe the great, and the beaver,[78] no less fine a spoil.[79] Costlier still, that ill-smelling plunderer of purses, the sable, wraps himself about the pleasure-glutted throats of princes.[80]

Throughout the womb of earth water is diffused, to create streams and rivers, marshes and lakes. The Euphrates flows through those lands where the great virago raised walls of clay for her Babylonian capitol.[81] The Tigris passes through the region of the earth where Crassus drank down gold, and Rome was exposed in his example. The Nile bears nourishing waters where you, great Pompey, confirmed how unsafe loyalty is when the prince is a boy.[82] Abana rushes forth, and Damascus waits for it to rise and nourish the fields with irrigating waters.[83] Shiloah, small but blessed, will behold a prophet, nay, God himself reshaping our existence. The Jordan is to be hallowed by the supreme honor of bathing the limbs of its noble creator. The Simois winds through the land of Troy—a happy land, had Paris loved more wisely. Sicily, fated to suffer under cruel tyrants, beholds the river Alpheus, and the stream of Arethusa. The Tiber, destined to possess the wealth of Rome and her crowning achievements, bears its waters angling toward the sea. The Po descends to the Ligurian plain, drives its waters along, and makes its way majestically toward Venice. Where the Rhone flows, Agauno saw her band of martyrs fight nobly until death.[84] Eridanus flows forth, the one stream which withstood the general disaster of Phaethon's flames.[85] The Seine wells forth where a warlike land has spawned great dynasties of rulers, the lines of Pippin and Charles. The Loire shimmers where the city of St. Martin lies between starry waters and brightly tinted fields.

Springs are wreathed with moss and riverbanks with turf. The field is clad with grasses, the grove with leaves. The plane tree flourishes on level ground, the alder on the slope, the sturdy box tree on rocky cliffs, the supple willow on the shore, the scented cypress on the mountain, the sacred vine on the hillside, the tree of Pallas in cultivated soil. There are the silvered poplar; the lotos, lover of the stream; [86] viburnum, suppler than its own shoots; the cornel, gnarled and bristling with long spears; the maple, hard and flexible, suited for strong bows; the holm oak, quivering with sharp and vibrant leaves; the yew, fell destroyer of the Cecropian bees; [87] the parent oak; the giant pine; the pigmy myrtle. The thorn tree, with armored body and menacing spines, the bristling bramble, and the buckthorn, similarly armed, fear nothing but the calloused hand. The beech, lover of ivy; the elm, bride of the vine; the wild briar, which scarcely separates itself from mother earth; the arching elder; the brittle sycamore; every kind of treee rejoices in the splendor of new foliage.

The grove of Alcinous, arising sometimes spontaneously, sometimes by the renewal of a parent stock, sometimes by a random seeding,[88] brings forth its fruits. Father Autumn smiles, and in his orchards, though but newly planted, the fruit comes all at once to a pleasing ripeness: the nut, clad in its jacket; the Carian fig, creased with wrinkles; that fig which was Adam's food; the pear, a common dessert; [89] the sorb-apple, a well-known cure for disorders of the stomach; [90] the white-skinned Syrian fig; the ruddy pomegranate; the hard-shelled chestnut; the downy-coated peach; the waxen sheen of the plum, which endures but a short time; the lofty winter oak, bowing to the earth; the pine, thrusting its top toward heaven; those nuts which were the food of scrimping Phyllis; [91] the cedar, bearing fruits endowed with a three-fold essence; the juniper, with polished berries of an almost peppery tang; [92] the almond tree, which is wont to issue forth in an early flowering; and whose topmost branches bear its nuts; the myrtle, dear to Venus; the laurel, sacred to Phoebus; and every tree that enjoys the distinction of a name.

Amid the flourishing wilderness the balsam, a noble tree, produces lowly foliage. Myrrh, too, by whose weeping over deceased bodies, lest they wholly dissolve, a second life is imparted. The sweet-scented frankincense spreads forth, which the Persian worshiper brought as a first offering to Christ. Cinnamon appears, sweet-flavored in its outer bark, but sweeter still within. The aloe, specially useful in the practice of medicine, seeps from its bark in amber-colored drops; thus too the tears of the Heliades, and cedar oil, that gum which Arabia sends, and that of the terebinth. Beneath the blazing sun the soil of India generates other essences, which create the atmosphere of ceremonial rites.

But still nearer to the dawn and the abode of Eurus, in the flowering bosom of the earth there lies a region upon which the sun, still mild at its first rising, shines lovingly; for its fire is in its first age, and has no power to harm. There a tempered heat and a favoring climate impregnate the soil with flowers and rich greenery. This little retreat harbors the scents, produces the species, contains the riches and delights of all regions of the world. In this soil ginger grows, and the taller galbanum; sweet thyme, with its companion valerian; acanthus, graced with the token of a perpetual blossom, and nard, redolent of the pleasing ointment which it bears. The crocus pales beside the purple hyacinth, and the scent of mace competes with the shoots of cassia. Amid the flourishing wilderness strays a winding stream, continually shifting its course; rippling over the roots of trees and agitated by pebbles, the swift water is borne murmuring along. In this well watered and richly colored retreat, I believe, the first man dwelt as a guest—but too brief a time for a guest.

Nature created this grove with affectionate care; elsewhere the wilderness sprang up at random.

The Aonian grove is born to be the delight of poets. Ida will provide ships' timber to bear away Paris' plunder. The glade of Aricia, now enervated by its trickling fountains,[93] puts forth leaves, and the Lycean grove on its grassy ridge. The grove of the Academy, surpassing in its charm, destined to harbor the high-sounding Sophists and Plato himself, comes into bloom. The Nine Sisters have abandoned the vault of the firmament, so great is the splendor of the Pierian grove. The shade of scented laurels about the Grynean temple is cherished by poets, and by the god of poets. India spawns trees which assail heaven with their tops; the Celtic lands, too, have their groves, and Sila, who raises her pine-covered summit to the stars, looking out upon the gleaming sails of two seas. Brittany has Broceliande, Touraine her Gatine, and Gaul has the forest of Ardennes.

Earth divides the jacketed vegetables into species: Italian chickpea and French bean; eyeless lentil, and peas which aid digestion; dark vetch and smooth kidney bean. Sparse winter grain grows hard, and ripe wheat swells; the slender oat grows tall, barley remains short-stalked.

As it slowly wakens to the presence of trickling water, the garden spreads forth such wealth as it possesses. Hyssop, a remedy for the lungs, dwells in the hollows of rocks, while open ground bears the drowsy poppy. Mustard takes command of uncleared fields, and chill Jove's-beard besieges lofty walls. Fountains possess the narcissus and

hedges the privet. Roses make the garden beautiful, lilies grace the
vale. Sleep-inducing lettuce [94] rises, together with scariola; low-lying
purslane, and endive, buried in the earth; the onion, well endowed
with coats; the harsh flavor of ligusticum; [95] garlic, and pot herbs,
which rejoice in a bed of turf; mint, diffusing its scent far and wide;
the streaked iris; the supple mallow, with the suppliant heliotrope;
lustful colewart [96] and obliging savory; satiricon, recalling old men to
youthful ways; [97] and she who is closed upon herself at close of day
and, at the day's reappearance, reappears herself, responding to the
sun like a bride. Artemisia, empowered to dispel those woes to which
women are subject, burgeons into bright foliage.[98] Sage, which im-
proves the flavor of festive meals, gives off its odor from its branches;
medicinal though it is, prodigal man, far gone in luxury, makes it a
dressing for his meat. Fennel appears, decked with delicate foliage,
and empowered to dispel the shade from clouded eyes: when the
snake casts off his skin, and with it the burden of age, it is with fennel that he restores his exhausted vision.[99] Marjoram takes root, and
wild thyme begins twining, an herb well known as an antidote to the
bite of serpents.[100] Calamint is the best pacifier for aching muscles
when rheumatism rages through the limbs. Pennyroyal is a proven
measure when there is suspicion of a poisoned draught, and chervil is
a no less efficacious herb. Wild nard, fumitory, and merry bugloss [101]
arise, to purge the effects of menstruation, spleen, and brain fever.
Summer savory, good for the kidneys, and the lesser plantain, with its
little pointed leaves,[102] which performs the same function even better;
rue, commended by Mithradates, and cabbage, the choice of Cato;
parsley, a crown fit for the locks of Hercules; [103] helvella, effective
against coughs; nettle, good for gout; camomile, bringing pleasant
rest to the sleepless brain; dittany, to draw out embedded objects;
panacea, known to bring the gift of healing to open wounds; anise
hot to the taste, which congests the genital ducts and closes the
broader canal to the flow of semen; [104] wormwood, which purges the
distempered stomach through perspiration; and violets, a prescription
against the heat. The seven kinds of tithymal arise,[105] which soothe a
troubled stomach—behold, O belly, the fights in store for you. Wild
cucumber appears, to create disruption in the stomach and grant purgation by its stormy effects. Socrates' hemlock springs up, together

with henbane, and hellebore, a plant intimate with death. In a new fit
of productivity the earth brings forth mandragora, prodigious plant,
to imitate our human countenance.

Proteus bears the various kinds of swimming life, and the scaly host
lays claim to its dominion: the whale, who frequents the coast of Brittany; the dolphin, with curved snout, who sports with the waves at
moments of impending danger; [106] the lamprey, highly valued as
food; the conger eel, prized still higher; the sea urchin, empowered to
cause fever; the foolish cod; [107] the pleasant-tasting dory, and the sea
pike, a fish of finer flavor; the mussel, and the conch tribe, possessed
of spacious palaces which are made anew as often as the moon renews her journey; [108] that Lethean fish who visits oblivion and self-
forgetfulness on anyone to whose line he attaches himself; the stickleback, whose wicked power makes old men itch with lust, and causes
members long dormant to rise again.[109] The Sirens, prodigies of the
deep, or at any rate many creatures of that order, dwell in hidden recesses of the sea.

Great numbers of fish, journeying abroad, move as guests through the
rivers and the realms of fresh water. In the river waters ocean-born fish
and those native to this region swim peacefully together: the angular
sturgeon, round mullet, spiny perch; the short roach, long barbel, broad
plaice; the rosy trout, tasty salmon, shad, surpassingly plump; the pike,
cruel tyrant and taskmaster. The carp lends distinction to the Loire,[110]
and there too swims the equally delicious maigre, a larger fish who
shuns the lakes and marshes.[111]

The fluid substance of water supports these kinds of life. That element touched by ethereal warmth bears many as well.

In the expanses of the air the winged creatures dance about, but a
good number never abandon familiar waters: the gull, which flies before the lunar surge of the incoming sea, and follows the waters as
they withdraw again; the densely feathered bittern and the long-
legged heron; the diver, glutted with fish, and the duck, foolishly
bold; the swan, who alone senses the terms on which life is lived and
goes forth singing in defiance of death.

The greater number are borne upward toward the heavens: the
phoenix, unique in its ability to renew itself by its own means; the
king of birds, who seized that youth who prepares the gifts of Bac-

chus and bore him through the night, himself a gift, to Jove; the falcon and the hawk, whose way of life is predatory, and who must subsist upon the host of lesser birds; the crane who describes in flight the shapes of written letters,[112] whenever he ventures forth from Strymonian waters; Juno's bird, a sport of nature; the dove, unsparing of her own breast; Philomela, who renews each spring her lament for her wrongs, and her sister, whose breast is stained with blood; the two cocks, that confined at home and that wild one which derives his name from Phasis, the land of Medea; the turtle dove, faithful lover; the bobtailed quail; the wily thrush, winning food by her charm; [113] Perdix—would that he had studied less and lived longer!—and the lark, who joyfully hails the new day; the sparrow, hopping hither and thither; the crow, who will behold ages to come; and the magpie, whose twofold coloring gives her a variegated appearance; the greedy vulture and the quarrelsome kite, deganerates both; the ostrich, who loves to make his home in solitary places; the finch, singing sweetly of tender love; the chatty parrot, who speaks with our voice; the raven, bird of prophecy, who does not recall nests of new offspring abandoned among the leafy branches,[114] the kingfisher and the woodpecker, guardians of shore and forest; the goose, who loves the open waters of the lake; the owl, whom the sun's kindly light makes blind, and the screech-owl, chanting doleful tidings in funereal tones.

When the feathered race had settled into these several forms, the birds differed greatly in bodily shape, inclination, and habitat.

## Chapter Four

Now since the fiery substance of the celestial sphere, due to that liveliness by which it was impelled, moved in a circular course around all this manifold progeny of creatures, it followed that the circumference of the firmament embraced within its vast compass the elements into which the universe had first been apportioned, and the portions of these portions.[115] For anything which is brought forth to assume the mode of being proper to its kind derives the causes and nature of its substantial existence from the celestial sphere, as though from a life-giving god.[116] For how are the stars borne about in a ceaseless journey, if not because they have imbibed ethereal nourishment? [117] How would the creatures of the land, the waters, the air, move if they had not received enlivening impulses from the firmament?

Ethereal fire, then, a lover and husband willingly drawn to the lap of earth, his bride, and there effecting the generation of all things, gives over that life which he has begotten by his heat to the nurture of the baser elements.[118] When the vital spirit of animate life has been summoned from the vault of heaven,[119] earth applies herself to providing nurture for bodily existences, and does not cease from the task of nourishment until she has ensured a sufficiency of created natures.

Thus Providence brought the creation of the universe full circle, from genus to species, from species to individual, from individual out again to first principles, in continual revolution.[120] When temporality was born from this first motion of incipient life,[121] the revolving of the firmament and the movement of the stars, number with its variations assumed control of the succession of the ages, which had had their beginning in the primal simplicity of eternity.

The totality of creatures, the universe, is never to be subjected to the infirmity of old age or sundered by ultimate destruction, for the basis of its survival consists in a maker and an efficient cause which are both eternal, and a material substance and form which exist together in perpetuity. For the primary substance, eternal permanence, simplicity fecund of plurality, one, unique, complete in and of itself, is the nature of God, whose infinitude of being and majesty no limit can circumscribe.[122] If in defining this mode of being you should call it "virtue" or "integrity" or "life" you would not be wrong.[123]

From this inaccessible light a radiant splendor shone forth—the image, or perhaps I may call it a face inscribed with the image,[124] of the Father. This is the wisdom of God, conceived and nourished by the living fountains of eternity. From this wisdom arises the deliberation, from deliberation the will, and from the divine will the shaping of cosmic life.[125]

Now the whole will of God consists in goodness.[126] Thus the divine will, or the goodness of the supreme Father, is the harmonious expression of His mind in uniform action. Who then would dare to disparage the universe and its eternal basis, when he may behold eternal causes working together to effect its moderation? From the will of God issues harmonious volition, from His wisdom deliberation, from His omnipotence both cause and effect.[127] From its participation in His stability and eternity the intellectual universe derives a prior knowledge of all that the sensible universe will gradually bring to realization through a carefully disposed series of secondary causes.

First there is Hyle; then comes the nature of the elemental qualities; the elements appear in response to this elementing nature, and elemented substances take their rise from the elements.[128] Thus the principles of existence depend on principles of their own, but these in turn depend on one sovereign principle. For if the firmament, and the movement of the stars, did not endow the elements with that capacity for change which they transmit, these would remain sluggish and inactive.[129] But the great lights, the sun and moon and all those which are called wandering spheres, and whose circling never ceases, do not suffer the elements of the underlying world to remain unmoving.[130] This elementing nature, then, is in fact the firmament, and those stars which traverse the circle of the Zodiac, for it is these that arouse the elements to their natural activity.[131] Thus these universal bonds do not grow weary and do not slacken, for they derive from one cardinal principle which binds them indissolubly together.[132]

The life and well-being of the universe depend on sovereign and ancient causes: spirit, sentience, a source of motivation, and a source of order. Noys and the divine exemplars live eternal; without their life the visible creation would not live everlastingly.[133] Hyle was in existence before it, preexistent in the substance and in the spirit of an eternal vitality.[134] For it is not to be believed that the wise creator of insensate matter did not first establish a basis for it in a living source.

The universe is an animal, and one may not detect the substance of animal life apart from the soul. Moreover, many things spring from the earth, but without the stimulus of a principle of growth neither tree nor shoot nor anything else would thrive. Thus from the life of the divine mind, from the spirit of Silva, from the world soul, from the growth-principle of created life, the eternity of the universe has its rise.[135] Knowledge reposes in God, in Noys; a rational plan exists in the firmament and intelligence in the stars.[136] And so in this great animal understanding and awareness thrive, and draw nourishment from their antecedent principles. The firmament learns from the divine mind, the stars from the firmament and the universe from the stars, whence their life derives and how they may discern the course of existence. For the universe is a continuum, a chain in which nothing is out of order or broken off. Thus roundness, the perfect form, determines its shape.[137] And so, although the flux of Silva often occurs, under the pressure of necessity, in a chaotic or violent way, that complex faculty or spirit which is present in the universe never permits the hostile force to overflow its bounds.[138]

Whatever exists in the dimension of time enjoys annual, or secular, or perpetual, or eternal existence.[139] The annual is dissolved by old age, the secular by the end of time itself. The perpetual vies with the eternal in endurance, but because at some time it had a beginning it may not attain the surpassing excellence of eternity. The universe sustains or prolongs the lives of its creatures, some in an annual, some in a secular, some in a perpetual state of activity. For the universe and time, owing to the principles from which they are sprung by a simultaneous act of creation, conform to closely related and virtually identical models.[140]

From the intellectual universe the sensible universe was born, perfect from perfect.[141] The creative model exists in fullness, and this fullness imparted itself to the creation. For just as the sensible universe participates in the flawlessness of its flawless model, and waxes beautiful by its beauty, so by its eternal exemplar it is made to endure eternally.

Setting out from eternity, time returns again to the bosom of eternity, wearied by its long journey.[142] From oneness it issues into number, and from the unmoving into movement. The instant present, the flowing away of the past, the anticipation of what is to come are the movements of time, and it moves through these channels in a perpetual ebb and flow. And though it shall have traveled these same roads over and over again in the course of eternity, still striving and forging ahead, it neither strays from the path nor turns back. And because the point at which its journeys end is the point from which they are renewed, it remains an open question whether the events of time past are not seen again in the same sequence.

By virtue of this necessity of returning upon itself, time may be seen to be rooted in eternity, and eternity to be expressed in time.[143] All that is moved is subject to time, but it is from eternity that all contained in the vastness of time is born, and into eternity that all is to be resolved.[144] Were it possible for time not to divide into quantity or issue into movement, then time would be identical with the eternal. Though it is variously named with reference to the course of the sun, it differs neither in its extent nor in its essential nature from the everlasting. Eternity, then, but also time, the image of eternity, shares the responsibility and labor of governing the universe. Eternity undertakes to impart life to the fiery bodies of the stars, and the purer substance of the ether. The activity of time sustains and gives rise to those subordinate existences which take their character from the lower atmosphere. Thus the universe is ordered by time, but time itself is governed by order.[145] For as Noys is forever pregnant of the divine will,[146] she in turn informs Endelechia with the images she conceives of the eternal patterns, Endelechia impresses them upon Nature, and Nature imparts to Imarmene what the well-being of the universe demands.[147] Endelechia supplies the substance of souls, and Nature is the artisan who compounds bodies,[148] the dwelling places of souls, out of the qualities and materials of the elements. Imarmene, who stands for temporal continuity,[149] in its aspect as a principle of order, disposes, joins together, and rejoins the universe of things thus comprised.[150]

# Microcosmos

*Chapter One*

When at last Providence was content with the eminently beautiful and surpassingly skillful arrangement of the sensible universe, she summoned Nature that she too might marvel and rejoice at that beautification which she had desired with all her heart. "Behold," said she, "the universe, O Nature, which I have brought forth from the primordial seed-bed, from the ancient turbulence, from the chaotic mass. Behold the universe, behold the exquisite subtlety of my work, the splendid construction, the grand display of created life, which I have made, shaped with zeal, and ingeniously conformed to its eternal idea, following as closely as possible my own thought. Behold the universe, whose life is Noys, whose form is ideal, whose substance is that of the elements. Behold: have I been attentive to your will in my labor? Do you greet with joyful prayers the birth of the universe?

"I need not mention with what a turmoil the roughness of Silva responded to my touch, and what diligence I brought to bear upon its reluctant unruliness until it grew tame under my shaping hand. I need not say how I chipped away the rust from the ancient elements, and endowed their reforged essences with the splendor that befits them. I need not explain how a sacred embrace brought together forces previously in conflict, and how a newborn mean rendered equal forces formerly imbalanced. I need not say how forms encountered substances, how it is that there is life on earth, in the sea, the air, the arching firmament. I would have you survey the heavens, inscribed with their manifold array of symbols, which I have set forth for learned eyes, like a book with its pages spread open, containing things to come in secret characters. I would have you regard the zones, and how, extending by fixed laws across the interpolar regions, they determine the climates of the underlying terrain. I would have you note the colural circles, and how, in their fourfold segmentation, they allow for the rotation of the firmament, but never make total that partitional course which they have begun. I would have you consider the Zodiac, which transcendent understanding has set atilt: hereby provision is made for the preservation of creation, which could not endure perpetually if the Zodiac always conducted the blazing sun in an unvarying course across the center of the earth. I would have you gaze on the Milky Way, moderating the cold of the northern regions; for to regions lying so distant, the heat of the sun does not bring its assuagement. And I would have you notice the line which corresponds to the two solstices, and likewise that which marks the meeting point of day and night at the time when they are equal.

I have fashioned the body of the sun with a brilliant, flaming, and rounded form. For him, set at the center of things, the planetary spheres join in harmonious chorus. The moon, placed at the boundary of the ether and the atmosphere, varies in power and appearance, and beholds the sun now from this side, now from that. I have set Venus in attendance on the sun, and assigned Mercury to travel close beside his light-bringing chariot. You observe that Jupiter moves in an extended circle, Mars in a more contained one, that the latter glows blood-red, and that the former is flattered by the glow of a friendly companion star. Saturn I have so far exalted that whatever be the elemental nature whose sign he occupies, to its influence he subjects the climate of the year.

"But why should I catalogue the positions of the heavenly bodies and the laws of the firmament when they are manifest on every side? You see how the earth, by a fertility derived from the elements, rejoices now in streams, now in meadows, now in bristling forests. Bounded by the domain of Amphitrite on all sides, it brings forth from within itself sustenance for living things. One region flourishes with fruits, another with trees, another with herbs and spices. One re-

gion teems with gems, another with different kinds of metal. In the
fluid element fish move about, even as the shapes of great beasts
roam through the realms of day. And lest the harmony of earthly life
be assailed by violent passions, to balance that heat which flows from
the sun, which the meridian bears along, I have poured out at the
center of earth a flood of water, the Mediterranean Sea. And because
I have divided up both this and the great Ocean into many sectors,
provision has been made for isolated regions, that the necessities of
life may reach them by ocean voyages.

"The many-colored host of the feathered race float along the stream
of the air, with all the freedom to journey where they will. I have
taught the winds to spread abroad the vessels of rain, that showers
of moisture may reagglomerate earth which has disintegrated into
sterile dust. I have sectioned off the vast ether by the imposition
of the zones, that the underlying earth may conform to their various
conditions."


*Chapter Two*
Now chaos had been divided into
parts, now Silva had been reborn to
her true beauty, and was a universe
worthy of the name. If her ancient origin intruded any trace of roughness, the shaping hand sought it out everywhere and banished it,
until, no longer resisting, Silva presented herself docile and well disposed to be wrought into the shapes of creatures.

"I set it to my praise and glory, O Nature, that I have so well cultivated my coarse materials. I have brought form to creatures and
yoked the elements by a harmony which has elicited peace and trust.
I have given a law to the stars, and ordered the planets always to
pursue the same undeviating course. I have curbed the sea with
boundaries, lest the land be flooded, and the earth rests, fixed by its
own weight, at the center of things. I have decreed that ethereal
warmth should bring forth vegetation, and that moisture should sustain what this ethereal warmth has produced; that the earth, loving
mother, should give birth to all things, and at their dissolution receive them back again into her tranquil depths; that every creature

should derive the seed and principle of its vitality from Endelechia, soul of the universe."

*Chapter Three*    "Thus, if you apply yourself and examine closely what I have done, all these are things whose shape, variety, stability, and arrangement you must surely admire. But since it befits the careful craftsman to make the final portions of his work a worthy consummation, I have decided to complete the success and glory of my creation with man.[1] I will bestow upon him abundant favor and abundant resources, that he may excel all my creatures by the privilege, as it were, the distinctive attribute of dignity.

"Now in harnessing the principles of things and sorting out the components of the resisting mass, I have applied my own strong hand against all unruliness, since necessity demanded it. But we have come now to man, and for his composition it seems good, and in no way unpleasant to me, that a fellow laborer should bring the aid of her enthusiasm to the task. Yet I recognize that the generation of the human soul, and the creation or instillation in this soul of the radiance of eternal vitality, are both tasks particularly demanding of my keenness.[2] For it is a task beyond your own understanding, Nature, and the faculties of any power you might muster, either to assess the soul's value or to express its majesty. My wish, then, is that you, Nature, by your own zeal and effort, seek out the dwelling places of Urania and Physis, both of whom are knowing and competent, both endowed with the power to perform the task in question. You will find Urania in close attendance upon my throne, and Physis sojourning among the lesser creatures."

At these words, Nature, her countenance equally expressive of gratitude and eagerness, hastened to obey the agreeable commands of Providence. For what gift could she have received more pleasing than the composition and formation of man, which she saw would be hastened to fulfillment by the summoning of these artisans? Accordingly she decided to seek out Urania first, since it seemed that she was the more distinguished, and her abode was near at hand. But though Nature had no doubt that she dwelt in the heavens, the vastness of this region seemed likely to produce a wandering and uncertain journey. For the circumference of the firmament is more vast than any breadth may define. Now since Urania must necessarily be concerned with the whole sky and all the stars, how was Nature to know which regions she might frequent in preference to the others? [3]

Anastros is a region of the firmament held to the norm of a uniform climate; its light unwavering, its calm perpetual, it is bounded by the purity of the ethereal realm, and closely resembles that realm in its climate.[4] Thus, as it is above the atmosphere in altitude, it is free from the passions of the atmosphere. It never grows heavy with rain, is never invaded by storms or disturbed by clouds. Here Nature sought Urania, if by chance the accessibility of the place or its aspect had attracted her when wandering at her leisure. But another region detained the queen of the heavens. Yet, however unrewarding to her prayers and useless to her undertaking, to have sought here was not a waste of time, for gazing on the magnificence of the region's glorious light made her tremble.

Setting out again on her long journey, she ascended to search the five parallel bands set between the poles of the firmament. But with their several extremes of climate, the midmost with its heat and the outermost with their cold, they promised as habitations unremitting hardship. And the two zones enclosed by the middle zone and the two extremes on either side vacillated between divergent climatic conditions. Through these spreading but distinct latitudes Nature traveled, searching with anxious care.

The colures part only to rejoin one another at a final meeting point through the concurrence of their arcs.[5] Among the colures as among the zones, though sought in both, Urania was to be found in neither.

Following the Milky Way like a highway, she encountered that radiance which its mass of stars produces through sheer multiplicity at the point where the Zodiac meets with the two tropics in its circular journey. Here she saw a numberless throng of souls clustered about the abode of Cancer. All these, it appeared, wore expressions fit for a funeral, and were shaken by weeping. Yes, they who were destined to descend, pure as they were, and simple, from splendor into shadow, from heaven to the kingdom of Pluto, from eternal life to that of the body, grew terrified at the clumsy and blind fleshly habitation which they saw prepared for them.[6] Some little time she spent pondering this spectacle, and what she sought was not to be found.

Accordingly she made her way along the line of the solstice, to that circle assigned to the stations and powers of the planets. The twelve sections of this circle are harder to traverse, as its slanted position retards progress. Leaving behind the orbits of the other planets, she entered the circle of the sun, for, since its course is less curving, it is subject to a stabilizing tendency. Looking down from this point at the top of the Zodiac, and surveying the universe, she gained no sight or sign, nor any information about her whom she sought. As a last resort she decided to visit the Aplanon,[7] highest and outermost limit of the firmament.

The ether and the bodies of all the stars are composed, not of the material elements, but of a fifth element, the highest in order, divine in kind, and of an invariable nature.[8] For if the firmament and the stars it bears derived their substance from the material elements, whose nature is changeable, they would reveal nothing certain, express no truth. This circle, then, outermost and all-encompassing, is neither fire nor derived from fire. Encompassing the spheres of the planets with its continuous and identical circlings, it sets these bodies in orbit by its own powerful motion.

In this, the all-forming region, a god of venerable aspect, and with the signs of the ravages of old age upon him, confronted her.[9] For the Usiarch here was that Genius devoted to the art and office of delineating and giving shape to the forms of things. For the whole appearance of things in the subordinate universe conforms to the heavens, whence it assumes its characteristics, and it is shaped to whatever image the motion of the heavens imparts.[10] For it is impossible that one form should be born identical with another at points separate in time and place. And so the Usiarch of that sphere which is called in Greek Pantomorphos, and in Latin Omniformis, composes and assigns the forms of all creatures.

"Hail, O Nature," said he. "You have come even to the summit of the firmament, and indeed you are worthy to be received in heaven, since you care for heavenly properties and essences with all the zeal of unflagging devotion. And that native ministress of the heavens whom you seek, Urania: behold her standing before you, gazing in wonder at the heavens,[11] calculating their recurrent motions and the periods of their orbits, in accordance with sure standards of observation."

Nature turned her dazzled and blinking eyes as best she could toward the thronging stars, and full into the impenetrable brilliance of the ether. Urania knew at a glance both who had come and her reason for coming. Cutting short the business of salutation, she forestalled Nature's attempt to speak by her divine insight.[12]

*Chapter Four*    "You bring, O Nature, the decrees of the most high God, and what the divine mind has even now willed to come about. God's will is that man be formed; his body will issue from the depths of chaos and his spirit from the powers above. Let the work be perfect,[13] let his beauty consist in the joining of his parts; it is God's will that nothing be lacking in his composition. It is God's will that the mixture be balanced, that balance effect a bond, that this divine bond bestow harmonious relation, lest it disgust the mind to dwell in shadowy blindness, and suffer the forced hospitality of the body, lest the spirit have cause to complain of the flesh that it is too much subject to its dictates, that this concord of unlike powers may come about peaceably, I am summoned to lend my help to the project. You are come to my dwelling no stranger, O Nature, for Noys bore me, too, your sister. You and I are of one kind; the universe at large is your dwelling place, mine this single portion of the ether.[14] My delight is in the heavens, in the stars, and I am reluctant to be drawn away from these things by my duties; a man of earthly nature is to be made, as a sojourner on earth, and the descent thither is not easy for me.[15] The dank impurity which adheres to the base earth will soon mar my brilliance. But since this creature is derived from archetypal patterns, it is useless to plead the journey as an excuse. In accordance with the sacred purpose of the divine mind I will execute the work of the office assigned me. And in following the form and order of the noblest model I shall pursue no rash or worthless end.

The human soul must be guided by me through all the realms of heaven, that it may have knowledge: [16] of the laws of the fates, and inexorable destiny, and the shiftings of unstable fortune; what occurrences are wholly open to the determination of will, what is subject to necessity, and what is subject to uncertain accident; how, by the power of memory, she may recall many of these things which she sees, being not wholly without recollection.[17] Let her align her genius and spirit with the gods and the heavens, that she may dwell as a queen within her earthly vessel. What virtue is in the stars, what power in the firmament, what vitality is in the poles of the heavens, what potency the rays of the two luminaries possess, and the five planets, these things let her know when she enters the vessel of the body. From the firmament let her learn a comely appearance, spiritual grace, and the laws of her behavior.

"By the laws of the firmament, man is assigned at birth his term of life and the means of its final disposition. Once having cast off the body he will come again to his native stars, one more divinity in the host of celestial powers. So shall it be, have faith; my voice is imbued with truth: for it is not permitted to the stars to lie. Go then, Nature, I follow; for no error can befall, if the way is determined by your guidance."

## Chapter Five

Nature was astounded by the divine interpretress, as she understood her exposition of their task and its principles and means of execution. And after receiving the whole speech without interrupting she expressed her approval of these declarations which so concurred with her own wishes by polite gestures and nods. That they might gain the express consent of the heavenly powers, and their guidance for the descent, now that the course of their solemn and sacred task had been determined, they entered the realm of pure and uncontaminated light,[18] far removed and wholly distinct from the physical world. Here, if you were to express it in theological formulation, is the secret abode of supreme and super-essential God. The heavens on either hand are inhibited by ethereal and divine powers;

their ordered ranks are themselves arranged in order, and each single power of the highest rank, the intermediate, or lowest, understands the principle of his assigned function, the value of his special task. For a single spirit pervades their several realms, conjoined with one another and arranged in unbroken succession, and imparts sufficient power to them all. But they do not receive a uniform power from this uniform spirit: those who are most nearly privy to the deliberations of the godhead are drawn, at times, when his will reveals itself openly and directly, even closer to the inner mind of God. Others, by virtue of being more distant, retain only a reduced and incomplete vision, and partake more sparingly of the sight of God and knowledge of the future.

From that realm where Tugaton,[19] the supreme divinity, has his dwelling place, a radiant splendor shines forth, nowhere partial, but everywhere infinite and eternal. This inaccessible brilliance so strikes the eyes of the beholder, so confounds his vision, that since the radiance shields itself by its very radiance, you may perceive that the splendor produces of itself an obscuring darkness. From this infinite and eternal splendor a second radiance distinguished itself, and from these two there arose a third.[20] These radiances, uniform and identical in brilliance, when they had together made all things bright, reabsorbed themselves again into the well of light which was their source.[21] To this threefold majesty, with a great outpouring of prayers, Urania and Nature together commended the purpose and success of their undertaking.[22]

Departing thence, they pursued their common path across the vast and spreading surface of the ether. Nor indeed could Nature have kept pace, had not Urania judiciously curbed her native and customary speed. These arcane regions once past, they approached that dividing line of the firmament where the ether commingles with the denser and less pure planetary heaven, imparting to it its properties. From the moment of their passage into this region, from the pure ether to dense air, from the temperate realm to the region of cold, Nature easily divined and understood the opposition which each met from the other.

Far below this level resided the Usiarch of Saturn, an old man everywhere condemned, savagely inclined to harsh and bloody acts of unfeeling and detestable malice. Whenever his most fertile wife had borne him sons, he had cut them off at the first budding of life, devouring them newly born. Ceaselessly on guard against childbirth, he neither paused for deliberation nor succumbed to pity, whereby he might sometimes have been sparing because of the sex or comeliness of the child. Nature was horrified by the old man's cruelty, and lest she should profane her divine gaze with so foul a sight, turned away her face in virginal alarm. One evil passion obsessed the old man, and he indulged in one form of savagery: he was still vigorous, and with a strength not yet impaired, and whenever there was no one whom he might devour, he would mow down with a blow of his sickle whatever was beautiful, whatever was flourishing. Just as he would not accept childbirth, so he forebade roses, lilies, and the other kinds of sweet-scented flowers to flourish. By the spectacle he presented he prefigured the hostility with which he was to menace the race of men to come by the poisonous and deadly propensities of his planet.

While Nature, after observing his labors, judged him cruel and treacherous, yet she believed the old man must be respected, inasmuch as it was said that Chronos was the son of eternity and the father of time. Neverthless, albeit their extended journey demanded that they stop and rest, it was agreed not to make this their resting place, where the peace of the firmament was so altered, and the air bristled with chill and icy harshness. Accordingly they mustered their courage, and crossing the barren and frozen wastes of Saturn they stopped at the abode of mild and beneficent Jupiter.

The Usiarch of this region is so propitious and well disposed that he is called in Latin Jove, from his power to aid,[23] and there is no belief more certain than that the effects of Jove's favor permeate every part of the universe. As they came into the pleasant atmosphere of this delightful orb, they beheld two vessels, placed at the very threshold of Jove's abode, one full of bitter absinthe, the other of sweet honey.[24] Souls were clustering about each vessel in turn, to taste of them, against the time when they should assume bodies. All life in the universe is subject to this condition imposed by Jove, that if any cause for joy arise from temporal life a cause for sorrow will occur as well. Seated in his council chamber, Jove shone in regal majesty, wielding in his right hand a scepter, and suspending from his left a scale, in the balance of which he determined the affairs, now of men, now of the higher powers. Whatever his most trustworthy balance had weighed with such measure, Clotho, a woman of commanding aspect, unfolded in due order through the process of time.[25] This woman, inasmuch as she measures out and sets in motion the precisely determined sequence of events, claims for herself a title of the utmost sovereignty. Thus all the area lying between the moon and Saturn is called the empire of Clotho. Reposing at this fine vantage point, Urania was grandly entertained and Nature still more so with this new and unusual view of creation. However, lest they delay too long in contemplating these novelties, they girded themselves for travel and resumed their laborious progress along their previous path.

As they approached the sphere of Mars, lying just below, with its curved but distorted bands, they were greeted by a murmur like that of water cascading down a steep slope. When, drawing closer, they could see clearly, Nature recognized by its seething and sulphurous waters the river Fiery Phlegethon, which issues from the sphere of Mars.[26] But at this time the fiery orb of Mars, emboldened by the favoring position of Scorpio as well as aroused by his own native propensities, shot out menacing beams into the fourth and seventh signs, and sought a way of breaking out of his orbit, so that, transformed to a comet, he might appear, blood-red and terrifying, with a starry mane. Quickening their pace, the goddesses hastened to pass by this realm, which seemed so ill-governed and teemed with hostile vapors, and proceeded to the dwelling of the life-giving sun.

Now the "Helian highway" in which the sun is borne on his annual journey was not everywhere the same, but consisted of four diversely colored segments. The first quarter of the circle flourished green as verdant Egypt with the buds of diverse flowers under the renewing influence of spring. The second, raging against spring's tenderness, with fiery vapors, grew dry and parched with the aridity of summer. The third presented an appearance compounded of the gold and green of autumnal ripeness. The fourth, too, extended through the breadth of three signs; on its surface shimmered a thin film of water, which the chill of winter had hardened into solid ice. That the single journey of this single power, so often reiterated, might present a more imposing spectacle, as he was borne through the fourfold change of

his orbit he underwent a fourfold change of countenance, passing from boyhood through pubescence into youth, from youth to manhood, from manhood gradually assuming hoary age through the interspersal of white hair. These various appearances the sun assumed as it was borne through the upper, lower, and intermediary stages of its journey around the slanting circle of the Zodiac.

Among the Usiarchs and genii of the heavens, whom eternal wisdom has appointed to adorn and govern the universe, the sun is preeminent in brilliance, foremost in power, supreme in majesty; [27] it is the mind of the universe,[28] the spark of perception in creatures, source of the power of the heavenly bodies and eye of the universe, and impenetrates all creation with an immensity of both radiance and warmth.[29] The instruments proper to the sun-god, his bow and lyre, hung close at hand, so that if he should in anger arm himself with his quiver, he might soothe and quiet himself by sounding the lyre. "Fruit of the Spring," a god of venerable aspect, and harmless Phaethon are both sons of the sun-god. Psyche and Swiftness are both Apollo's daughters.[30] On the right hand the youths and on the left the maidens stood about the light-bearing chariot. Psyche was gathering from her father's burning lamp those beams which he was to spread through the heavens and the earth.[31] Swiftness, following always in the path of the sun, governed the movement of time in this way: a revolution of the firmament simultaneously determined the extent of a single day; the spans of the months were resolved by the phases of the moon; the year's cycle by the number of the months; and by the accumulation of years the procession of the ages.

Urania, recognizing here an office like her own, would willingly have remained with these kindred maidens, were it not that this would have involved a delay of the task at hand. Indeed, as they had spent so long in admiring contemplation of the sun, Nature suggested that charming Lucifer, and its companion Cyllenius should be visited without loss of time. Accordingly they entered—for it would have been wrong to pass them by—the circles of Mercury and Venus, intricately connected with one another and with the sun.[32] And had not Urania noted carefully the junctures and points of intersection, they would have been carried by a roundabout path back to the sun whence they had come.

Mercury travels around the orbit of the sun on a closely contiguous path, and thus is often heralded by the very power whose herald he is. Because of the law which governs his orbit he rises at times above the sun, and sometimes lurks beneath him. Compliant and indecisive, Mercury does not point to the coming of misfortune in the affairs which he governs by his stellar clarity. Rather his relations with other powers vindicate or corrupt him. Joined with the madness of Mars or the liberality of Jove, he determines his own activity by the character of his partner. Epicene and sexually promiscuous in his general behavior, he has learned to create hermaphrodites of bicorporeal shape.[33] This god held a slender wand in his hand, and his feet were winged, lightly shod and bound, as befitted one who performed the office of interpreter and messenger of the gods.

Venus, while she touches the orbits of Mercury and the sun at certain points, encompasses both with the fullness of her own. Maintaining a median climate between the extremes of heat and moisture, she draws forth by the largesse of her own nature the fruits of budding plants, and inspires the renewal of all creatures by her generative impulses. Lending her authority to the evidence of favorable stars, she brings to a more tender fruition those births over which she presides. Astrologers believe that whatever incites the human longing for pleasure becomes vehement through the influence of Venus' star. The radiant countenance of Venus gives great delight to her beholders. Her distinguishing ornament is a torch, now smouldering, now bursting into flame. Little Cupid clung to her left breast.

Passing between those spheres which are far apart at one moment and impinge closely upon one another the next, Urania and Nature, after looking back along their route as far as the ultimate void, discussed the complex order which they had seen. Since an easy and sloping path led down to the circle of the moon, the lowest of all, it seemed no long journey. They made haste unawares, and arrived at their goal more swiftly than they had expected. Here a boundary was interposed between the ether and the atmosphere. The separation of the natures of the two regions was preserved by the mediating power of the moon.[34] Above was endless calm, perpetual quiet, the unbroken peace of the ethereal regions. Hence the higher powers, as they are not driven into one state and another by the process of change, are in no way diverted from their inviolability and intrinsic dignity. The penetrating mind of Greece concluded that in this region of the heavens, since its nature is unchanging, soothing, and quiet, were placed the Elysian Fields,[35] where blessed souls were enshrined in a gracious and hallowed state, never to be deprived of light. In the regions below were disposed the more turbid properties of the atmosphere, whose volatile appearance is altered whenever some chance occurrence offers to its passions the material they demand. So mankind, inhabiting this unquiet region, the very image of the ancient chaos, must needs be subject to the force of its upheavals. For the necessity which arises from the fluctuating state of Silva, and which has been rigorously purged from the heavens and the stars, remains fully active in the lower regions. Thus the moon, traveling her divisory and lowly course, is crude and heavy of body by comparison with the other spheres.[36] Feeding upon the divine and immortal vitality of the heavenly fires,[37] she transmits even that Aethericon,[38] which is the essence of bodily growth, to the lower world. Her gleaming body, which shines with the reflection of the sun's brilliance, is consumed and restored by a regular and unvarying flow of substance. Because her radiance is produced by the radiance of another, Ptolemy of Memphis named her the planet of the sun.[39] Just as she both arouses and allays the swellings of the ocean, so she is recognized to influence earthly existence more powerfully in proportion as she is closer to the earth. Constantly passing through the same course of withdrawal and return, with ceaseless and unwearying speed, she can claim the most immediate power over the affairs and destiny of humankind. Her divinity, one and the same beneath a variety of powers and duties, manifests itself, now as Lucina in her radiance, now as the Huntress with her quiver, and now as Hecate with garlanded head.[40]

## Chapter Six

The part of their task now accomplished and the hardships overcome on their journey gave the goddesses cause for encouragement. They had traversed the seven planets, reigning in their spheres, and the realm of the stars. Their talk as they went was of the old man's curved scythe, and the helmeted form of the bloodthirsty warrior, and Jove between the two; the sun and moon, illuminating the universe; and the two who are bound to one another, Mercury to Venus, Venus to Mercury. It was pleasant to wonder at the heavenly domain of the unfathomable mind, the intricate work of the almighty: where the colures lie like a girdle across the plane of the sky; the binding zones; the stability of the poles; how the bending Zodiac and the Milky Way, obliquely crossing the sky, are firmly set in their places; how the sun is content to move between its tropics, and never exceeds these duly established limits; how its orbit carries it by a slanting path, so that it does not shine with equal radiance on all the lands over which it passes; how the sign of the Ram balances the periods of night and day, and determines the equinox; that it is the sun's task to banish the shades with daylight, to restore color to earthly life, and splendor to the sky; that the moon, eager follower and handmaid of the sun, has power over earthly life even as over the tides. Bearing witness to its creator, the supremely beautiful frame of creation pleases them by its form and its components. Whatever the stars, whose condition is nobler because of the ennobling influence of heaven, release into earthly life degenerates little by little as it descends; thus all that lies beneath our turbulent sky is to be regarded with mistrust by virtue of its imperfect and fluctuating condition.

Pausing in this region of the heavens, Nature directed her clear gaze upon new sights.

*Chapter Seven*  Now along this lunar boundary, which is, as it were, the midpoint of Homer's "Golden Chain," [41] and the umbilicus which unites the higher with the lower universe, a crowd of thousands of spirits thronged joyously together, like citizens of a crowded city.[42] As Nature stared with fixed, unblinking eyes at their numbers and their diverse shapes, Urania spoke: Learn—for it is not fitting that Nature, so diligent a seeker of causes, should be in doubt —, learn, I say, O Nature, what spirits these are, what their distin-

guishing qualities, and in what diverse stations they serve the almighty.

"The firmament, the ether, the atmosphere, the earth: this fourfold
sphere includes the whole content of the universe. The firmament is
uniform, of a common density throughout, unchanging in climate.
You have come to know the twofold division of the ether, and the atmosphere, and the threefold partition of the earth. Each of these
realms and each of their subdivisions has its spiritual realm, each has
its angels.

"The firmament is alone filled by God's presence; [43] for it is un-
thinkable that the sacred and unchanging godhead should establish
its dwelling place upon the lowly elements, the impure earth, and the
brawling atmosphere. God the disposer of all regards the lower and
lesser universe from a seat on the celestial heights. Albeit He is not
seen directly, withdrawn behind the darkness which surrounds His
divine majesty, yet He shines manifest in the visible evidences of His
handiwork. The angels are His creation, whom He has so ordered
that by a system of unbroken continuity the highest are linked with
the intermediate and they in turn with the lowest.[44] His are the animate powers of the firmament, the celestial fires, rational creatures of
a kind neither destroyed by death nor altered by passion. The intelli-
gent race of man, whom we have undertaken to fashion, will claim, as
a worthy place of habitation, the lowest sphere, the earth. In addition, a third race occupies the middle realm of the universe, combin-
ing the attributes and participating in the condition of the two
extreme orders. For multitudes of the angelic host share the divinity
of the stars, in that they do not die. They share the nature of man in
that they are impelled by the effects of passion.

On a still loftier height, if there be any place higher than the firmament, Tugaton, the supreme divinity, has his dwelling. Those gleaming hosts whose nature is blazing fire, who are both created and
named "burning spirits," [45] stand about him and approach him closely.
Because of their constant and unwearied inclination toward God,[46]
beings of this order are deemed "limbs" or parts of the divine by a
sort of similitude: for in truth there is no division in the godhead.
Owing to their proximity, these beings are continually drawn close to
God and receive the hidden deliberations of His mind upon things to

come, which determine what will come to pass as the destiny of the universe at large, by inevitable necessity, through the agency of the lower orders of cosmic spirits.[47] Enjoying the vision of eternal bliss, delivered from all the vexations of distracting concerns, they repose in the peace of God which is beyond understanding.[48]

This host, then, pure, wise, and obedient to God, dwells from the eighth sphere of the firmament, and the brink of the pure ether, down to the sphere of the sun. Along the line of descent leading to the moon others are interspersed, who are inferior to the majestic powers above, through a certain diminution of brilliance and power, in proportion as their place is lower. All the same, spirits of this rank are blest with understanding and recollection, and their powers of vision are so subtle and penetrating that, plumbing the dark depths of the spirit, they perceive the hidden thoughts of the mind.[49] They are wholly bound to charity and the common good, for they report the needs of man to God, and return the gifts of God's kindness to men, and so seek to show at once obedience to heaven and diligence in the cause of man.[50] Thus the name "angel" denotes their office, not their nature. Accordingly, when the new design, the new creation of man has taken place, a "genius" will be assigned to watch over him,[51] drawn from this most merciful and serviceable race of spiritual powers, whose benevolence is so deep-seated, and unalterable, that they shun, out of a hatred of evil, any contact with the vile or displeasing; but when, through the inspiration of divine powers, some virtuous act is undertaken, they are ever at hand.[52]

"In the sublunar atmosphere the higher portion differs from the lower more by its climate than by its location. The highest levels are more refined, and somewhat warmer, being affected by the contingency of the fiery condition of the ether, so far as mean things can be influenced by great, or sluggish life by that which is more highly animated. The class of spirits who dwell in the atmosphere, but in serenity, maintain calm of mind, as they live in calm.[53] Second in rank to these is the genius which is joined to man from the first stages of his conception, and shows him, by forebodings of mind, dreams, or portentous displays of external signs, the dangers to be avoided.[54] The divinity of these beings is not wholly simple or pure, for it is enclosed in a body, albeit an ethereal one. For the creator drew forth the distilled essence of ethereal calm and ethereal fluidity, and adapted divine souls to a material which was, so to speak, unmixed. Since their bodies are virtually incorporeal, and subtler than those of lower creatures, though coarser than those of higher powers, the feeble perception of man is unable to apprehend them.[55]

"Below the midpoint of the teeming air wander evil spirits, and agents of the lord of cruelty. They cannot avoid the taint of earthly foulness, for they hover close to the surface of earth. These beings, having been only slightly cleansed of the ancient evil of matter, have been contained within extremely narrow bonds by the great foresight of God. And since they persist in wickedness and the desire to do harm, they are often empowered by divine decree to inflict torment on those stained with crime. Often, too, they decide for themselves, and inflict injury of their own accord. Often they insinuate themselves invisibly into minds at rest, or concerned with their own thoughts, through the power of suggestion. Often, assuming bodily existence, they assume the forms of the dead.

"The first rank of spirits I call the guardians, those intermediary the interpreters, and the lowest the renegade angels. Consider now those earthly beings who inhabit the world.[56] Wherever earth is most delightful, rejoicing in green hill, flowery mountainside, and river, or clothed in woodland greenery, there Silvans, Pans, and Nerei, who know only innocence, draw out the term of their long life. Their bodies are of elemental purity: yet these too succumb at last, in the season of their dissolution.

"The Plutonian Usiarch, whom I may call Summanus,[57] or lord of the shades, is preeminent in his influence from the limits of the atmosphere down to the surface of earth, and the empire over which he rules begins at the circle of the moon. But I pray you, let not a power whose potency is limited to the atmosphere appear to your judgment as vile or unworthy of the respect due to majesty. The atmosphere is the means of breathing, and without the gift of the atmosphere the health of created life cannot endure.

"Such, then, is the multitude whom you behold spread forth; who never abandon their positions, above or below the sphere of the moon, nor cease to perform the tasks assigned them."

*Chapter Eight*[58]

"Observe, O keen of mind, how the world is formed, and by what means the elements are interwoven; with what zeal fostering Noys dispersed the rude heap, that they might preserve a stable and unbroken harmony; what binds extremes to medians, what establishes the bonds between them, why the firmament moves, and why the earth remains in place; the motions of the planets, what power is assigned to each, what rising and setting, what location, motion, and laws; why the rotation of the Aplanon runs counter to the seven planets, and their wandering course through the signs; why delight is in the power of Venus, wars belong to Mars, cold to Saturn, and fair weather to Jove; what are the powers of the sun, Mercury, the moon, at what pace they move through their signs and orbits: should you investigate heavenly number, or meditate the astrologer's art, it is this which shows what the course of fate will grant or deny.

"You have learned why winter fetters the earth, and spring sets it free, why the summer is hot, why autumn is the season of harvest; why Boreas is chill and Favonius mild; why the one lays waste the flowers, while the other adorns the earth; how, when Jove descends into the lap of his spouse, the world is renewed, and all the earth swells with new life; that Ceres may discover the lost daughter whom she had sought with the torch, and that, creeping slowly upward, the maiden may show forth her head; that birds may be engendered in the forest, fish in the seas, that all the fields may flower, and all the groves put forth new leaves.

"You have learned what harmonious proportion unites souls to bodily members, so that a single bond of love links unlike natures, though the flesh be of earth and the mind ethereal; though the one is gross, the other volatile, one dull, the other keen; it is thus that the simplicity of the soul enters the condition of otherness,[59] and its single substance is divided among diverse kinds of life. But what is thus held by the fetters, locked in the prison of the body, and lies, all but buried beneath its burden, will return to the glory of its birth, the kingdom of the father, if it is wise, and does not submit to the tyranny of the flesh.[60]

"Now, unfolding still higher things through philosophy, behold what is permitted to death, what is death's cause and who its author, by what authority it draws down, in what maelstrom engulfs everything that air, earth, or water sustains. Yet if a mind which consorts with truth may inspire true understanding, death deprives a thing of its form, but does not steal away the essence of that thing. For the subject matter remains the same, though its form pass away, and a new form only gives this matter a new name.[61] Form flows away, the essence of the thing remains; the power of death destroys nothing,[62] but only disunites united parts.

"Learn through philosophy what is pleasing in itself, what must be pursued for another reason, what is fitting, and what is not. Be zealous to distinguish justice from injustice, truth from falsehood, and to test your observation by reason."

*Chapter Nine*   During this conversation they had come to that region of the atmosphere disputed by the sons of Aeolus,[63] freezing at one moment and burning the next, often assailed by showers of hail, and as often resounding with thunder clouds. Urania was amazed at this irregularity, having no experience of variability, or of anything divergent from calm. She saw that the fluid substances of the elements, owing to some mutable property, were altered by every incursion of contrariety: that they were now assailed by rain clouds drawn up from the sea, now burdened with clouds of a denser kind which the earth produced. Insofar as all these were at odds with her sense of order, they were painful to her spirit, and vexing to behold. Recoiling in horror from the inconstancy intrinsic to this region, and having passed through the interstices of the elemental structure, they hastened forward and came to rest in the lap of the flourishing earth.

Granusion is located in a secluded and remote spot at the eastern limit of the earth. Having received, through the tender ministry of the sun in its freshest youth, a most happy evenness of climate, it is lush with grass and burgeoning with rich growth. The name of the place is

Granusion because it is continually bringing plants of all sorts to maturity.[64] This hidden valley presents to the various uses of mankind whatever allays disease, whatever is conducive to health, whatever pleasurably arouses sensual desire, be it tree, grass, or herb. This is the one privileged spot in all the world, I think, which feels none of the effects of elemental strife, and is capable of enjoying a full and perfect evenness of climate. The place maintains perpetual springtime, for the variations of winter, summer, and autumn are kept at bay by the favor of heaven. Yet it knows not that this is really the gift of God.

Since she thought that Physis might be found in this very place, Nature suggested that they turn aside. Spontaneously, the radiant beauty of the place took on a fuller luster, for it had received a premonition that Nature, mother of generation, was at hand. The earth, through that fecundity which it had received out of the womb of Nature, suddenly teemed with life, and its hidden powers put forth clusters of sturdy new growth. The grove of the Heliades emitted a still more abundant essence; boughs of Sabaean fragrance vied in exuding their richness. Nearby, groves of balsam and cinnamon gave forth their scents, one faint, the other filling the air. Thus everything to which the delightful east gives birth and nurture rose up at Nature's arrival with a certain festive air. There was a stream, descending to the floor of the glade from a spring on the steep, not in such a way as to strike the ear with a tumultuous roar, but rather soothing with its gentle murmur. Charming as it was to hear, it was more charming still to behold; its descending flow was of an ethereal purity, almost as if it had cast off its bodily state and regained the status of a pure element. It descended in bends and windings in order to provide a sufficiency of moisture for all the verdure of the glade. A wood, which surrounded the place and its activity on every side, served the twofold purpose of shielding it from the hot sun and preventing open access. Within these confines an ethereal heat operated in the moist soil, so that at one place an array of flowers, at another herbs, at still another spices sprang up almost unbidden.

Looking down they saw Physis seated there, attended by Theory and Practice, her daughters and inseparable companions. She was occupied with the study of created life, in this calm and sequestered place, where nothing could disturb her. She had taken as the subject of her thought the origins of all natural things, their properties, powers, and functions, and the whole range of the Aristotelian categories. By an unwavering path from those principles established by divine wisdom through genera, species, and individuals she followed Nature, and whatever is included under that name. Whenever her thoughts descended from the heavens and heavenly bodies, she sought to explain by their composition the behavior of animate creatures: that the timidity of the hare was due to coldness, the boldness of the lion to fire, that cunning was instilled in the fox and sloth in the donkey by the phlegmatic humor in one case and the melancholic in the other. She observed that the state of an animate body fluctuated under the shifting influence of mutability; thence arise the afflictions of disease, first troubling the spirit, causing its bodily dwelling to totter, and finally casting it forth from its home altogether. She labored to oppose this by seeking out remedies by whose tempering influence imbalance might be restrained when it sought to do harm. Indeed she accomplished this with such penetrating insight that the very elements, primary components of the universe, and the components of these components, served the purposes of her work.[65] Not content with herbs, plants, and grasses, she wrung curative effects even from metals and stones. It must be added to the sum of her wisdom that by artful mixing she could use even the deadliest poisons most effectively in the work of healing.[66]

Physis sat dreaming, and was absorbed in deducing, from the potentiality of Nature, and in a highly imaginary way, the composition of man, when a ray of light gave notice of the approach of Urania, and by reflection in the nearby spring revealed the countenance of the still absent goddess.[67] Theory, the first to recognize the two guests, awoke her mother and bade her sister rise. Embraces were exchanged and they greeted one another by name. That reverence was observed which the entertainment of guests demands. Once invited to seat themselves, they explained briefly why they had come, with no interruption hindering their discourse.

And behold, Noys appeared before them, and, having appealed for silence, began to speak.

*Chapter Ten*

"Goddesses, my beloved children, whom I created before the creation of time; even I take pride in my progeny. This is the final object of my will; you have come to the final stage of my plan and enterprise. If there is any lack among the creatures and forms of the universe, your hands, inspired by me, may make it good. To the extent that what I have wrought is less than complete, less than perfect, less than beautiful, it seems vile to me. That this sensible universe, the image of an ideal model,[68] may be able to attain fullness in every part, man must be made, his form closely akin to the divine, a reverend and blessed conclusion of my work. Like all in the universe which draws life from the eternal ruler he will be a worthy and in nowise ignoble representation of my wisdom. He will derive his understanding from heaven, his body from the elements, so that while his body sojourns on earth his mind may dwell far above.[69] His mind and body, though of diverse natures, will be joined into one, such that a mysterious union will render the work harmonious. He shall be both divine and earthly, comprehend the universe about him through knowledge, and commune in worship with the gods. Thus he will be able to conform to his two natures, and remain in harmony with the dual principles of his existence. That he may at once cherish things divine and have charge of earthly life, and satisfy the demands of a nature which is drawn to both, he will possess the gift of reason in common with higher powers: and indeed only a thin line will separate him from these powers. Brute beasts plainly reveal the grossness of their faculties, their heads cast down, their gaze fixed on the earth; but man alone, his stature bearing witness to the majesty of his mind, will lift up his noble head toward the stars, that he may employ the laws of the spheres and their unalterable courses as a pattern for his own course of life.[70] The heavenly powers, the stars, the firmament, will speak to him, and Lachesis reveal to him her deliberations. He shall behold clearly principles shrouded in darkness, so that Nature may keep nothing undisclosed.[71] He will survey the aerial realms, the shadowy stillness of Dis, the vault of heaven, the breadth of the earth, the depths of the sea. He will perceive whence things change, why the summer swelters, autumn blights the land, spring is balmy, winter cold. He will see why the sun is radiant, and the moon, why the earth trembles and the ocean swells. Why the summer day draws out its long hours, and night is reduced to a brief interval. It is my will that the elements be his, that fire grow hot for him, the sun shine, the earth be fruitful, the sea ebb and flow; that the earth give nourishment to its fruits, the sea to its fish, the mountains to their flocks, and the wilderness to its beasts for him. I have established him as ruler and high priest of creation, that he may subordinate all to himself, rule on earth and govern the universe. But when at last the tottering structure of his bodily dwelling falls, its binding harmony dissolved,[72] man will ascend the heavens, no longer an unacknowledged guest, to assume the place assigned him among the stars." [73]

*Chapter Eleven*    "This task imposes an obligation on each of you, for it is threefold: the composition of a soul from Endelechia and the edifying power of the virtues; the composition of a body by the conditioning of matter; and the formative uniting of the two, soul and body, through emulation of the order of the heavens. The first task plainly belongs to Urania, the second to Physis, the third, O Nature, to you. While much has been given you for the fulfillment of your task, yet much has been withheld; it is left for you to undertake, as you see fit, the engendering of a human soul from the already created Endelechia, and a body of those elements extracted from the mass of chaos.[74] Are you aware of what zeal, what perseverance, what effort the moulding of these demands? It is, indeed, a weighty and intricate task, involving most difficult calculations, that I have imposed. And although, as often happens, your memory will falter in the face of the many details, the weight and seriousness of the task, recourse may be sought to these aids which I shall provide: I bestow upon Urania the Mirror of Providence, upon Nature the Table of Destiny, and upon you, Physis, the Book of Memory.[75] This threefold gift embodies, to speak plainly, insight into the deliberations of God; true knowledge; and the most productive kind of certitude."

The Mirror of Providence was of vast circumference and boundless breadth, its surface extending forever, its shining glass such that whatever reflections it had once received no rubbing might erase, nor age make faint, nor destruction mar. There lived ideas and exemplars, not born in time and destined not to pass away in time. This Mirror of Providence is the eternal mind, in which resides that unfathomable understanding, that intellect which is the creator and the destroyer of all things. Among the exemplars might be discovered the model of anything, of whatever sort, and its quality and quantity, and when and how it had come to be. Here was Silva, still hidden by the darkness of her ancient condition; now she assumed the shapes of new creations, through the refining work of God; now came the balancing and self-containing harmony of the elements, containing and interweaving them with each other. Now appeared the rounded and revolving vastness of the firmament, now emerged the vital fire, Endelechia,[76] embracing the whole, within and without. Now the fiery bodies of the sky, now those orbs jointly empowered to provide for the world, the vivifying and generative sun, and the moon, assisting all birth. Now the planets and their signs. Now the races of land-going, swimming, and feathered creatures, as the friendly quality of this or that element received their several species. This array of different forms posed a grave difficulty to Urania, and she spent a long and anxious time before she could be sure that she had found the reflection of the image of man.

The Table of Destiny, too, was of vast extent, but finite. Its substance, neither smooth nor shiny, was of the cruder material, wood. Here, in much the same manner as in the Mirror, appeared the shape of all creation, tinted with the hues of life. The difference between Mirror and Table was that the Mirror was particularly concerned with the unchanged state of heavenly natures, while the Table for the most part exhibited such products of the temporal order as were subject to change. Thence it is that Atropos, Clotho, and Lachesis, sisters obedient to Providence and fate, are assigned to keep a common careful watch, though in separate realms, over the workings of the universe.[77] Atropos governs the sphere of the firmament, Clotho the planets' wandering,[78] Lachesis the affairs of earth. Thus the Table of Destiny is nothing else but the sequence of those things which come to pass by the decrees of fate. Here too there are traces of the divine handiwork, but they are slight. Natural events and those taking place in time were contained here in extensive array. Herein were shown the causes of the ancient strife of matter, and the miracle of the creator whereby peace was found amid so vast a disunity of ingredients. Here was shown how species and form come together with substance, and the marvelous means whereby they receive the impress of divine ideas. Here too how species and virtue descend from the firmament to earthly life; and what the efficacious motions of the stars portend; how the substance of dead creatures is restored to life, and how a seed-plot is established for the universe where it sprouts forth and grows afresh. Here were every animal, all species and all natures. And thus, when the mother of generation had studied carefully the fatal Table, humankind, lurking among so many species, could barely be discovered.

A long chain of fate and history stemmed from the first man, who was shown as distinguished by the elevation of his head. Now were shown the ways of fortune, the downtrodden simplicity of the masses, the venerable exaltation of kings. Now poverty begot misery, or over-abundance led to dissipation. Most people preserved a median existence between the two extremes. Now a human life was ordained to the toils of war, the pursuit of wisdom, or some other kind of endeavor. The sequence of the ages, introduced by the pure primal state of the Golden Age, could be seen degenerating little by little, to end at last in an age of iron.

There remained the Book of Memory, written not in ordinary letters, but rather in signs and symbols, its contents brief and compressed into a few scant pages. In this brief compass the combined workings of Providence and fate could be deduced, and partially understood, but they could not be foreseen. For the Book of Memory is nothing else but the intellect applying itself to the study of creation, and committing to memory its reasoning, based often upon fact, but more often upon probable conjecture. Here appeared, though not with the same clarity, the same accounts of created natures that had been shown before. Here, however, much fuller and more careful information was given regarding those creatures which are beheld in bodily form. Here the four components of the world's body were shown, summoned from their natural litigation to hear the eternal judgment. The plan was shown whereby love and compatibility were imparted, so that component parts might form a compound body, and sundered plurality commit itself to unity. There was shown the close kinship within the watery and feathered races, though distributed into species distinct in property and shape. Here appeared the plan whereby Nature provides scales for one creature, feathers for another, whereby birds possess a language of sweet song, while fish remain forever dumb. Here appeared the various four-footed creatures accustomed to a domestic state, and others whose wild character was given to savagery. The plan appeared whereby fierceness is common to lions and boars, while in the deer and the hare fiery vitality grows weak. Here were careful observations concerning the powers of herbs. The plan was shown whereby the potency of one is in the seed, another's in the stem, that of a third in the root. Here was shown the beginning of all that generation draws forth to substantial existence and the passing away of those things whose substance corruption destroys. Amid so great a host of earthly natures Physis discovered only by great effort the image of man, faintly inscribed at the very end of the final page.

*Chapter Twelve*    The goddesses quickly took in all that had received ordered expression in these exemplars of creation. Nature, the first to gird herself for work, summoned her sisters and fellow laborers. Urania applied hands well skilled in art and mustered her faculties. But Physis, though outwardly uncomplaining, was hard put to contain the silent anger in her mind. The task was one for intellect, and it seemed that any lesser power must be overborne by the weight of the undertaking. "It is an easy task to fashion brute beasts," she thought, "involving little grace and nothing whatsoever of art. But man, the second universe, demands the understanding and care of a nobler and abler power. Man, God's true likeness, a spark drawn from the heavens, must be the product of both art and intellect."

She was mistrustful of the coarse substance of the elements, and doubted that they would prove wholly adaptable, for she saw in them the stains and the ineradicable evil of Silva. The violent and teeming state of matter in its primordial confusion terrified the prospective artisan: fire warred with moisture, and moisture with fire, and they adopted one another's roles. It was her task to recall to order whatever through too much force had transgressed its bounds. The malignity of bodily nature and its teeming components might well make her fearful lest their instability should scorn and repudiate all form, and mock her discipline. Perhaps the tumultuous mass might be subjected to harmony, but this too promised a grim struggle. The universe, complete and with fully developed powers, contains throughout its being the seeds of its continuance. But mortal man is not so made, and he whose task it was to create the lesser world of man would have to construct it very differently; for the universe is a work of perpetuity, man of time; there is one breath of life in the universe, another in man. This was a task for intellect, for the fire of a keen mind, and a hand capable of so edifying man that he, who is not preserved by his relations with external life, might survive through a power within himself. She saw that the materials of the work and their fundamental principles, and hence the composite substance of man, were governed by the stars and the firmament, and influenced by the changing state of the powerful moon.[79] Moreover, certain evils appear in bodily life from the very moment of birth; and Physis feared them. Two aspects of the undertaking made her blush with confusion, and Urania undertook these: to expel the evil taint from Silva, and to contain fluid matter within fixed bounds.

The human race, although, being mortal, it is inhibited by its condition, must yet be so reformed that it may rise to dwell among the heavenly powers, and to subject to its laws whatever the star-bearing sphere impels by its circling, and by all these means redeem the taint of its earthly beginnings and its innate evil. Physis returned to herself, knitted her brow, and lo, entered upon her work more carefully.

*Chapter Thirteen*    Now there were two principles of created life, unity and diversity.[80] Diversity was of extreme antiquity. Unity had no beginning, but was simple, inviolate, remote, complete in and of itself, infinite, and eternal. Unity is God, and diversity nothing else but Hyle, and that which lacks form. All-producing divinity refined diversity, limited its boundlessness, shaped its formlessness, disentangled its involvement, imposing upon Hyle the definition of the elements, defining those elements by essences, essences by qualities, and assigning essence and quality to matter.[81] Thus matter, though tainted by the innate contagion of Silva, was transformed through the conversion of the elements to an essentially substantial condition, and assumed a bodily state.[82]

Full of purpose, Physis subdued these essentially bodily ingredients, the four primary foundations of her creation, the sinews and nerves of her work. So long as she adhered to the pattern of the supreme examples of the works of divine wisdom, she traveled a smooth and open path, for the diligence of God the creator had gone before, and taught Silva already to submit and obey. It had recalled the unruly from tumult, the seething from strife, it had adorned the unrefined with splendor. It had established a continuity among parts whose conditions were incompatible in many ways, establishing laws and unbreakable bonds to ensure peace. Lest any room be afforded to random error, it had established certain dwellings for the individual elements. There was not the least particle anywhere which might oppose itself to her control.

Nonetheless, the rough necessity of ever-flowing Silva lurked close beneath the surface. The fluctuating mass harbored an evil tendency to injure or destroy the glory of the divine handiwork. Physis kept watch, that she might avert, if possible, or at least contain any treachery of this sort. Yet at the very point when she was disciplining the material for her destined task, what had flowed into her hands flowed away again, and the shape she had sought to fashion dissolved. Physis was appalled at this inconsistency which hindered her progress, and cursed the unbridled lawlessness of her material. She took pains and labored mightily, so far as Nature would admit, to check its fluctuation and contain its flowing away. Moreover, she realized that not the elements themselves but mere fragments of the elements had been given her to build with, scraps and leavings which she had found discarded after the completion of the universe, and that it was beyond even a skilled craftsman to perform her task or bring her work to completion with such imperfect materials. Physis applied her keen understanding to these great and pressing difficulties. As so much was lacking she decided to reexamine the material which provided the basis of her work. In this she beheld only images of the elements, and not their true nature in the integrity of its purer substance: not those elements which achieve perfection, but the dregs of the essences of the elements, gross remnants of their original simplicity.[83] Examining the fiery, earthy, and other bodies with her own cruder understanding, she conjectured that their powers would become full and entire insofar as they were so themselves. Drawing forth the qualities of moist, dry, warm, chill from the material substrate which presented itself to her scrutiny, she first separated them individually and then blended together those closely related—for such had been the great design of the creator.[84] She separated, that is, those properties which had been simple in their substantial nature. She mingled those closely related that they might form compound substances. To consider fire and water: if dryness or moisture is associated with the heat of the one, or with the chill of the other, a certain kinship is revealed by their alignment and fusion. In the case of air and earth: if heat or cold is associated with the moistness of the one or the dryness of the other, such a relationship is called a mixture. Physis worked with the utmost care at this mixing of qualities, for this aspect of her task was clearly no idle matter. She recalled that in the human anatomy certain parts were to be formed from simple, others from compound materials. And when the elements adapted themselves to one another on the basis of closely related properties, that coherence appeared which in other cases is called complexion.[85] Physis applied these elementary complexions to the human constitution in such a way that a man's basic nature would conform to the principles from which it arose. Melancholy and phlegm are the result of earthly gravity, on the one hand, and watery instability on the other. Choler rages like fire, a sanguine disposition is airy and mild. Moreover, that carefulness on the part of Nature which is reflected in man is not found in other animals. For an imbalanced mixture of humors all too often leads to a distorted complexion in brute beasts. The donkey is made stupid by phlegm, the lion wrathful by choler; the dog is wholly pervaded by his aerial sense of smell. The human condition is utterly unique. A balance is created among qualities and quantities by the mingling of the humors. Human nature has been wrought with all possible care into a whole, such that paucity and excess present scarcely any threat to the soundness of the work.[86] For it would have been improper for the future abode of intellect and reason to suffer imbalance or disruption through any uncertainty in its design.

Thus, when the humors had been set in balance and their powers adjusted, when their qualities had combined to produce the essential substance of bodily existence, that fullness of parts ensued which constitutes the body. And since this fully realized solidity demanded the further development of form, Physis first carefully divided the bodily material into three portions. These she gave the rough outline of lineaments, and soon shaped them into the form of members. The first she called the head, the second the breast, and the third the loins, according to the properties she observed in them. These three in particular of the body's many parts, these narrow chambers out of its general extensiveness she chose to receive the brain, the heart, and the liver, the three foundations of its life. Physis knew that she would not go astray in creating the lesser universe of man if she took as her example the pattern of the greater universe.[87] In the intricate structure of the world's body, the firmament holds the preeminent position. The earth is at the lowest point, the air spread between. From the firmament the godhead rules and disposes all things. The powers who have their homes in the ether and the atmosphere carry out its commands, and the affairs of the earth below are governed by them. No less care is taken in the case of man, that the soul should govern in the head, the vital force established in the breast obey its commands, and the lower parts, the loins and those organs placed below them, submit to rule. So Physis, skilled artist as she was, prepared the brain as the future seat of the soul, the heart as the source of vitality, and the liver as the source of appetite, and took pains to prepare a divine dwelling place for divine guests. For she gave a rounded shape to the head, which occupied the chief position, following the example of the firmament and the sphere of the heavens.[88] She raised the head to the position of a temple or capitol for the body as a whole,[89] making it stand out toward the heavens; for it was fitting that she so exalt the region of the head, where the divine quality of pure reason was to dwell. She placed this noblest part of the body, charged with the duty of understanding, furthest away from the cruder organs of digestion, lest its perceptions be affected by that waste which comes from the digestion of food.[90] So too, following an inscrutable design, she encased the substance of the brain, which she had made soft and fluid, with a hard earthen casing. She had decided to employ a soft and transparent material in creating the brain, so that the images of things might impress themselves more easily upon it. Then, dividing the whole cavity of the skull into three chambers, she assigned these to the three functions of the soul.[91] In the frontal chamber provision was made for imagination to receive the shapes of things, and transmit all that it beheld to the reason. Memory's chamber was set at the very back of the head, lest, dwelling at the threshold of perception, she should be troubled by a continual invasion of images. Reason dwelt between these two, to impose its firm judgment on the workings of the others. She also set the organs of sensory perception close about the palace of the head,[92] that judging intellect might maintain close contact with the messenger senses.[93] For as these are prone to many sorts of error, it is best that they should not be far removed from the seat of wisdom. Physis hastened to make the head a home ideally suited, with its several chambers, for the soul and its vital powers, that she might proceed to its accessory parts.

Now while the operative power of the pure soul is itself one and pure, it is not uniform in operation. For sight is transmitted through the eyes, and hearing through the ears. Similarly it adapts itself to the structural differences of the other organs which it employs. Thus, though they stem from a single source, the senses perform a variety of functions. One perceives colors, another sound, one flavors, another odors. Touch is found spread abroad throughout the body. The senses possess a power of apprehension imparted by the elements from which they are derived. Sight was not formed without fire, nor smell and hearing without air. Taste, affined to water, and touch, similarly related to earth, operate on kindred materials. Whatever be the elemental quality by which they subsist, it is by this same quality, and by distinguishing similarity, that they proceed.

*Chapter Fourteen*

Man was formed with masterly and prudent skill, the masterwork of powerful Nature. Belief holds that wisdom chose the head as its seat, and divided it into three chambers. In these three is placed the threefold power of the soul; each part fulfills its function in an unalterable sequence. The recollective faculty is placed at the rear, the speculative power is foremost, and reason exercises its power at the center. All share in the work when the five attendant senses inform them of the external events which they perceive. A messenger·of sense enters and arouses the tranquil mind to confirm the matter by sure judgment.

That nerve which illumines the eyes with its light draws from the brain the power which it radiates.[94] An inner light, a daylight of the soul, responds to the rays of the sun's fire and the brilliance of the ether. From this cooperation the power and the organ of sight derive the principle and the material means of their existence. A beam of this inner light applies itself to the forms of things, and makes a careful record of them.[95] However, it does not perceive all things with the same clarity; its power is feeble at one moment, ample at another. It fixes most clearly upon things which are bright and most like itself, but applies only vaguely and dully to things unlike. As splendor is at home with splendor, so is light with light; in shadow and darkness sight falls idle.[96] The smooth surface of a rounded and polished body: this was the form best suited to the eyes.[97] The images of things adhere most clearly to a smooth surface, and it will possess a capacity for livelier motion. Because the eye is endowed with both motion and brightness both are reflected in the creation of its form. That it may not suffer injury its light lies well covered, contained by a sevenfold jacket.[98] The forest of the brow protects it from any hazard which so delicate and unstable an organ might fear. It is not for nothing that there are two, since if one fails the other may perform the work of its partner.[99] The lids are a bed of rest, at the time when peaceful sleep soothes the laboring orbs. Just as the sun, the world's eye, excels its companion stars and claims as its own all below the firmament, even so the sight overshadows the other senses in glory; the whole man is expressed in this sole light. To one who asked why he was alive, Empedocles replied, "That I may behold the stars; take away the firmament, I will be nothing." [100] The unseeing hand spoils its work, the foot strays drunkenly, when they perform their tasks in darkness, without light.

The hearing holds an inferior place, is inferior in power, more sluggish in perception, and of less usefulness. Sound emerges from the windpipe and stirs the still air. Once aroused, the agitation spreads, until the last wave of motion slackens, having attained its limit and been drawn out to its full extent. Air provides the substance, and the instrument of speech the form; from these two sound derives the shape and essence of speech. For the tongue forges sounds to the form and image of speech and serves as the hammer in the process. Shaped by its efforts, the articulated substance of speech travels to the open ears. Having first been admitted to the ear as though to the outer vestibule, the voice calls out and is admitted to the inner rooms. The ear keeps outside the rhythm and the resonance of the words, but the thought signified gains admittance. The ear interprets what comes from without, the tongue reveals what is within; and each requires the aid of the other. The channel of the ear is tortuous, lest cold air should pass by too open a path to the brain.[101] Nature feared for its frail condition, and so a winding path leads inward from the curving shore. All that Rome learned, all that you studied, O Athenians, whatever the east possessed of Chaldean wisdom, whatever Aristotle perceived in his inspired breast, or the Pythagorean band, or the Platonists, whatever Gaul debates in syllogism with subtle speech, or Italy pronounces in the art of medicine, all this is the result of hearing. The wise and learned letter would perish if man existed with deaf ears.

Though the tongue promotes and assists the needy Arts, it is well known for doing harm in countless evil ways. Whenever it whispers indiscreetly in the jealous ear backbiting and poisonous words, it separates loving brothers, destroys friendships, breaks bonds of trust, divides marriages, makes the land teem with thieves, the forum with quarrels, the city with war; it discovers secrets, and opposes all established institutions.

The taste is derived from water, and its affinity is with fluids. Whatever imparts a taste to the mouth contains moisture. Nature assigned this delicate sense to the delicate palate, and labors to provide it with agreeable food. I could wish poor man still more deficient in this sense, that he might taste more things with his mind, less with his mouth.[102] It is the taste which seeks to strip the earth of game, the air of fowl, the sea of its fish. Voluptuous hunger, whenever it forsakes the halls of the great, is ruinous to those of meager means. It squanders money, wantonly consumes wealth, and feasts on the inheritance which a better sense had amassed and retained.

Physis divided the Panchaean or scenting power of smell between the nostrils.[103] This sense both undermines and stimulates the work of the brain. Tainted air is the cause of the odors of things, and the work of the nose is impaired by this impurity.

The touch is yet more sluggish than the sense which judges food, or that which comes into play when a scent is drawn in. Touch campaigns in bed, serves the cause of tender love, and is given to exploring slily the smooth belly below the tender breast, or the soft virginal thigh.

The heart is second in dignity to the brain, though it imparts to the brain the source of its vitality.[104] It is the animating spark of the body, nurse of its life, the creative principle and harmonizing bond of the senses; the central link in the human structure, the terminus of the veins, root of the nerves, and controller of the arteries, mainstay of our nature, king, governor, creator.[105] It is a noble lord journeying abroad through all the state of the body, to the limbs and the ministering senses, each of whom it maintains in the function assigned to it. Its sacred shrine is within the breast, its royal palace and imperial throne; its form is such as its brother element, fire, provides, whose crown tapers upward to a point. A thing too intensely hot can be injured by its own fire; when the heart burns, the moist lungs bring their aid. Soft and airy, they surround and support the heart, ministering to its heat with their cool moisture. For wrath dwells in the kindled heart, finding there a congenial dwelling, a home well adapted to its habits,[106] until some trivial event irritate the heart; then fever is spread abroad into all parts of the body. Nor, once stricken, is it wont to languish, but almost anticipates the pain it is to suffer, and succumbs to the evil in advance.

A moist principle created the brain, a fiery one the heart, and an aerial blood compounded of the two the tender liver. Porous and hollow in form, it is bordered by the diaphragm and the spleen on one side, the stomach on the other. The liver receives at last what the hand places in the mouth and the teeth grind up and the stomach digests. It adapts ànd transmutes the food, imbuing it with the appearance of blood, and produces the nourishing fluids of the body. Flowing thence in all directions, the veins bear the appropriate portion to each member. Then when the fluids have again been cleansed and purged it chooses among them and rejects the good in favor of better. It assigns to the spleen the task of cleansing the blood, and thereby atoning for its vicious and foul actions, so that the sensual appetites of its master are served only by healthy fluids and pure blood. For this reason, though the stomach is the vessel of food and gall its cook, the liver is finally the overseer of the process; for natural desire dwells in the region of the liver, and love is indeed a fell tyrant over our flesh.

The lower body ends in the wanton loins, and the private parts lie hidden away in this remote region. Their exercise will be enjoyable and profitable, so long as the time, the manner, and the extent are suitable.[107] Lest earthly life pass away, and the process of generation be cut off, and material existence, dissolved, return to primordial chaos, propagation was made the charge of two genii,[108] and the act itself assigned to twin brothers. They fight unconquered against death with their life-giving weapons, renew our nature, and perpetuate our kind.[109] They will not allow what is perishable to perish, nor what dies to be wholly owed to death, nor mankind to wither utterly at the root. The phallus wars against Lachesis and carefully rejoins the vital threads severed by the hands of the Fates. Blood sent forth from the seat of the brain flows down to the loins, bearing the image of the shining sperm.[110] Artful Nature molds and shapes the fluid, that in conceiving it may reproduce the forms of ancestors.[111]

The nature of the universe outlives itself, for it flows back into itself, and so survives and is nourished by its very flowing away.[112] For whatever is lost only merges again with the sum of things, and that it may die perpetually, never dies wholly. But man, ever liable to affliction by forces far less harmonious, passes wholly out of existence with the failure of his body. Unable to sustain himself, and wanting nourishment from without, he exhausts his life, and a day reduces him to nothing.

In creating man Physis had to bestow limbs of which the universe has no need: [113] eyes to keep watch in the head, ears for sound, feet to bear him, and all-capable hands.

# Notes

## Notes to Introduction

1. On the "discovery" of self-consciousness in the twelfth century see Marc Bloch, *Feudal Society*, tr. L. A. Manyon (Chicago, 1961), pp. 106–8; Jacques Le Goff, "Métier et profession d'aprés les manuels de confesseurs au moyen-âge," in *Beiträge zum Berufsbewusstsein des mittelalterlichen Menschen*, ed. Paul Wilpert (*Miscellanea medievalia*, 3, Berlin, 1964), pp. 51–55. The emergence of the intellectual is discussed by Le Goff, *Les intéllectuels au moyen âge* (Paris, 1957), pp. 9–59; R. R. Bolgar, *The Classical Heritage* (Cambridge, 1954), pp. 130–201. On the new status of technology, see esp. Olaf Pederson, "Du Quadrivium à la Physique," in *Artes liberales: von der Antiken Bildung zur Wissenschaft des Mittelalters*, ed. Josef Koch (Leiden and Cologne, 1959), pp. 107–23; also Maurice de Gandillac, "Place et signification de la technique dans le monde médiéval," in *Tecnica e casistica* (Rome, 1964), pp. 269–73.

2. "Du Quadrivium à la Physique," p. 109.

3. See the fundamental study of M. D. Chenu, *La théologie au douzième siècle* (Paris, 1957), pp. 323–50.

4. See Wolfram von den Steinen, *Der Kosmos des Mittelalters* (2d ed., Bern, 1967), pp. 253–79; R. W. Southern, *Medieval Humanism* (London, 1970), pp. 33–38; Robert Javelet, *Image et ressemblance au douzième siècle de St. Anselme à Alain de Lille* (Paris, 1967), I, 390–408.

5. See Philippe Delhaye, *Le Microcosmos de Godefroy de St. Victor: Étude théologique* (Lille and Gembloux, 1951), pp. 137–44.

6. See the discussion of von den Steinen, *Der Kosmos des Mittelalters*, pp. 245–52.

7. Otto von Simson, *The Gothic Cathedral* (New York, 1956), pp. 21–58; Le Goff, *La civilization de l'Occident médiéval* (Paris, 1967), pp. 407–12.

8. *Der Kosmos des Mittelalters*, p. 249.

9. The title *De mundi universitate* assigned to the work by its editors is evidently drawn from Bernardus' description of its theme in his dedicatory epistle (see below, Dedication, n. 2). *Cosmographia* is the title most commonly assigned in the mss., though it is possible that this too is largely generic in reference; the two books of the work are often cited directly by their own names, *Megacosmus* and *Microcosmus*. On the medieval history of the terms *cosmographia, megacosmus* (or *macrocosmus*), and *microcosmus*, see Walter Kranz, *Kosmos* (*Archiv für Begriffsgeschichte*, 2, Bonn, 1955), pp. 131, 136–37.

10. On the development of this theme in the twelfth century see Delhaye, *Le Microcosmos de Godefroy de St. Victor*, pp. 137–77; Javelet, *Image et ressemblance*, I, 230–32; Chenu, *La théologie au douzième siècle*, pp. 34–43; Brian Stock, *Myth and Science in the Twelfth Century: A Study of Bernard Silvester* (Princeton, 1972), pp. 197–207, 275–76. On the broader history of the theme, Rudolph Allers, "Microcosmus from Anaximandros to Paracelsus," *Traditio*, 2 (1944), 319–407; Kranz, *Kosmos*, pp. 115–74.

11. "The Fabulous Cosmogony of Bernardus Silvestris," *Modern Philology*, 46 (1948–49), 92–116.

12. See especially his *Platonismo medievale: studi e ricerche* (Rome, 1958), pp. 122–50.

13. See below, Introduction, Section 6.

14. See Jean Györy, "Le cosmos, un songe," *Annales Universitatis Scientiarum Budapestensis: Sectio Philologica*, 4 (1963), 87–110; Hennig Brinkmann, "Wege der epischen Dichtung im Mittelalter," *Archiv für das Studium der neueren Sprachen*, 200 (1963–64), 419–31; Leo Pollman, *Chretien von Troyes und der Conte del Graal* (Tubingen, 1965); *Das Epos in den romanischen Literaturen* (Stuttgart, 1966), pp. 59–88.

15. Alfred Adler, "The *Roman de Thebes*, a 'Consolatio Philosophiae,'" *Romanische Forschungen*, 72 (1960), 258.

16. Edmond Faral, "Le Roman de la Rose et la pensée française au xiiie siècle," *Revue des deux mondes*, 35 (1926), 430–57; Helen Waddell, *The Wandering Scholars* (7th ed., London, 1934), pp. 115–22.

17. See Friedrich von Bezold, *Das Fortleben der antiken Götter im mittelalterlichen Humanismus* (Bonn, 1922), pp. 79–84; Wolfram von den Steinen, "Les sujets d'inspiration chez les poètes latins du xiie siècle," *Cahiers de civilization médiévale*, 9 (1966), 373–83.

18. Isidore of Seville, *Traité de la nature (De naturis rerum)*, ed. Jacques Fontaine (Bordeaux, 1960), pp. 11–13.

19. "L'idea di natura nella filosofia medievale prima dell' ingresso della fisica di Aristotele: il secolo xii," in *La filosofia della natura nel Medioevo: Atti del Terzo Congresso Internazionale di Filosofia Medievale, 1964* (Milan, 1966), pp. 34–35. This article (*in toto*, pp. 27–65) is an excellent

summary of and bibliographical guide to scholarship in this area, and I have drawn on it at several points in this section of my introduction.

20. See the discussion of Hans Liebeschütz in *The Cambridge History of Later Greek and Early Medieval Philosophy*, ed. A. H. Armstrong (Cambridge, 1967), pp. 576–78.

21. See Pierre Courcelle, "Etude critique sur les commentaires de la Consolation de Boèce," *Archives d'histoire doctrinale et littéraire du moyen âge*, 12 (1939), 12–76; *La consolation de philosophie dans la tradition littéraire* (Paris, 1967), pp. 29–66; Gregory, *Platonismo medievale*, pp. 1–15.

22. *Opusculum* 22, *Patrologia latina* (hereafter cited as *PL*), 155, 170. On this work see Gregory, *Platonismo medievale*, pp. 17–30; Eugenio Garin, *Studi sul platonismo medievale* (Florence, 1958), pp. 23–33.

23. *De divina omnipotentia* 3, *PL*, 145, 600–1; cp. Manegold, *Opusculum* 14, *PL*, 155, 163.

24. See P. O. Kristeller, "The School of Salerno: Its Development and Its Contribution to the History of Learning," in his *Studies in Renaissance Thought and Letters* (Rome, 1956), pp. 495–551.

25. On the integration of medicine with physical science in general, see Kristeller, "The School of Salerno," p. 515, and his "Beiträg der Schule von Salerno zur Entwicklung der Scholastischen Wissenschaft im 12. Jahrhundert," in *Artes liberales* (cited above, n. 1), pp. 87–88, which provides manuscript evidence for the incorporation of medicine into the Liberal Arts curriculum by the mid-twelfth century.

26. See Pederson, "Du Quadrivium à la Physique," pp. 113–14.

27. On the relation of old and new in Guillaume's thought, see Southern, *Medieval Humanism*, pp. 79–81, who points out that medical literature was his one truly new resource; Peter Dronke, "New Approaches to the School of Chartres," *Anuario de estudios medievales*, 6 (1969), 124–25, 128–32. On the influence of medicine on his thought, see Richard McKeon, "Medicine and Philosophy in the Eleventh and Twelfth Centuries: the Problem of Elements," *The Thomist*, 24 (1961), 231–243; Heinrich Schipperges, "Einflüsse arabischer Medizin auf die Mikrokosmosliteratur des 12. Jahrhunderts," in *Antike und Orient im Mittelalter: Vorträge der Kölner Mediaevistangun 1956–59*, ed. Paul Wilpert (*Miscellanea Medievalia* 2, Berlin, 1962), pp. 138–45; Silverstein, "Guillaume de Conches and Nemesius of Emessa: On the Sources of the New Science of the Twelfth Century," in *Harry Austryn Wolfson Jubilee Volume* (Jerusalem, 1965) II, 719–34.

As Stock shows, *Myth and Science*, pp. 59–62, the serious concern with myth and the application of the *trivium* to philosophy typical of Guillaume was being rejected in his own day by more rigorously rationalistic scholars like Hermann of Carinthia, who emulated the Arabs' concern with the *quadrivium*.

28. *Opusculum 22, PL,* 155, 171.

29. *De philosophia mundi* 1. 23, *PL,* 56. On Guillaume's rationalism see Schipperges, "Einflüsse arabischer Medizin," pp. 138–39, 144–45.

30. *Glosae super Platonem,* ed. Edouard Jeauneau (Paris, 1965), pp. 211–15.

31. On this tendency see Jeauneau, "Macrobe, source du platonisme chartrain," *Studi medievali,* 1 (1960), 12–15.

32. On this subject, which has yet to be adequately explored, see Hans Liebeschütz, *Medieval Humanism in the Life and Writings of John of Salisbury* (London, 1951), pp. 76–78; Gregory, "L'idea di natura," pp. 52–62.

33. On this theme see Gabriel Nuchelmans, "Philologie et son mariage avec Mercure jusqu'à la fin du xiie siècle," *Latomus,* 16 (1957), 84–100.

34. See the references gathered by Gregory, "L'idea della natura," pp. 41–42; Pederson, "Du Quadrivium à la Physique," pp. 109–10; Stock, p. 61.

35. *De eodem et diverso,* ed. Hans Willner, *Beiträge zur Geschichte der Philosophie des Mittelalters,* 4 (1903), No. 1, p. 10.

36. As Gregory notes ("L'idea di natura," p. 42), such ideas are, for Adelhard, virtually synonymous with *ratio.*

37. *De eodem et diverso,* p. 32.

38. The source of this conception is Macrobius, *Commentarii in Somnium Scipionis* 1. 14. 15, ed. James Willis (Leipzig, 1963), p. 58. Cp. *Cosmographia* 2. 8; ed. C. S. Barach, J. Wrobel, *De mundi universitate libri duo sive megacosmus et microcosmus* (Innsbruck, 1876), p. 47, lines 1–3.

All subsequent references to the *Cosmographia* will be to book and section of the work itself, and to page and line in the Barach-Wrobel edition (BW).

39. See Gregory, *Anima mundi: la filosofia di Guglielmo di Conches e la Scuola di Chartres* (Florence, 1955), pp. 155–74; Garin, *Studi,* pp. 62–68. On the broader question of the theme of the world soul in the early Middle Ages see Allers, "Microcosmus," pp. 355–63; Gregory, *Platonismo medievale,* pp. 122–50.

40. See Gregory, *Platonismo medievale,* pp. 135–37; "L'idea di natura," pp. 44–46.

41. See Richard Lemay, *Abu Ma'shar and Latin Aristotelianism in the Twelfth Century* (Beirut, 1962).

42. *Platonismo medievale,* pp. 122–50; "L'idea di natura," pp. 42–50.

43. *De sex dierum operibus,* ed. N. M. Haring, "The Creation and Creator of the World according to Thierry of Chartres and Clarenbaldus of Arras," *Archives d'histoire,* 30 (1955), 195–96.

44. *Didascalicon* 1. 10, ed. Charles H. Buttimer (Washington, 1939), p. 18.

45. *De essentiis* 1, ed. P. Manuel Alonso (*Miscellanea Comillas,* 5, Santander, 1946), p. 62.

46. See especially the *De processione mundi*, ed. Georg Bulow, *Beiträge*, 24, No. 3 (1925), and Gregory, "L'idea di natura," pp. 48–49.

47. *De septem septenis* 7, *PL* 199, 961–62.

48. In an anonymous commentary, probably of the twelfth century, on Boethius' *De consolatione philosophiae*, the *Timaeus* is described as "liber quem fecit [Plato] de natura animae": *Saeculi noni auctoris in Boethii Consolationem Philosophiae commentarius*, ed. E. T. Silk (Rome, 1935), p. 155. (All subsequent references to this work will be to "Silk.")

49. See Chenu, *La théologie au douzième siècle*, pp. 19–23.

50. See Philippe Delhaye, "*Grammatica* et *ethica* au xiie siècle," *Recherches de théologie ancienne et médiévale*, 25 (1958), esp. pp. 91–110.

51. *Prologus in Eptateuchon*, ed. Edouard Jeauneau, *Medieval Studies*, 16 (1954), 174.

52. *Saturnalia* 5. 1. 18–19, ed. James Willis (Leipzig, 1963), p. 243.

53. *Mathesis* 1. 4, ed. W. Kroll, F. Skutsch (Leipzig, 1897), I, 11–15.

54. *De institutione arithmetica* 1. 1, ed. G. Friedlein (Leipzig, 1867), pp. 9–10.

55. *Commentarii in Somnium Scipionis* 1. 2. 17–18, p. 7.

56. "De libris quos legere solebam," in *Poetae latini aevi carolini*, I, ed. Ernst Dümmler (Berlin, 1881), p. 543.

57. *De scriptoribus ecclesiasticis* 28, *PL*, 160, 554.

58. Commentary on Boethius, *De consolatione philosophiae* 3, metr. 9, *PL*, 64, 1241.

59. *Introductio ad theologiam* 1. 20, *PL*, 178, 1029. See also the references gathered by Richard McKeon, "Poetry and Philosophy in the Twelfth Century: the Renaissance of Rhetoric," in *Critics and Criticism, Ancient and Modern*, ed. R. S. Crane (Chicago, 1952), pp. 300–1.

60. On this term see Edouard Jeauneau, "L'usage de la notion d'integumentum à travers les gloses de Guillaume de Conches," *Archives d'histoire*, 32 (1957), 35–42.

61. Gloss on Boethius, *De consolatione* 3, metr. 12, quoted from Ms. Troyes 1331, f. 69r by Jeauneau, "L'usage de la notion d'integumentum," p. 45.

62. See Henri de Lubac, *Exégèse médiévale* (Paris, 1959–64), II. ii, 189–200; Jeauneau, "L'usage de la notion d'integumentum," pp. 35–42. As Stock shows, *Myth and Science*, pp. 54–59, Guillaume de Conches, the most systematic of the analysts of mythical *integumenta*, was clearly conscious of the analogy of his procedures with biblical exegesis.

There is, however, an essential decorum in twelfth-century interpretations of the *integumenta* of the *auctores;* they do not admit a truly spiritual meaning, or exceed what can be imagined as the allusive range of the author's conscious verbal artistry. Thus they are really different in kind from such later treatments of classical myth and poetry as those of Bersuire and the English "classicizing friars," in whose work, as Judson B. Allen remarks, "exegesis and arts commentary come together." See his lucid discussion, *The Friar as Critic* (Nashville, 1971), pp. 18–30.

63. See Chenu, *La théologie au douzième siècle*, pp. 129–31, 174–78, 289–308; Roger Baron, *Science et sagesse chez Hugues de St. Victor* (Paris, 1957), pp. 148–66.

64. "Image de Dieu et nature au xiie siècle," in *La filosofia della natura nel Medioevo* (cited above, n.19), p. 287.

65. *La théologie au douzième siècle*, pp. 178–90.

66. See *Didascalicon* 3. 4, ed. Buttimer, p. 54, and de Lubac, *Exégèse médiévale*, II. i, 291–94.

67. See *Expositio in Hierarchiam Coelestem* 1. 1; *In Ecclesiasten Homiliae*, 5, 10; *PL*, 175, 926, 156, 177.

68. *Didascalicon* 1. 2, ed. Buttimer, pp. 5–6.

69. *De Sacramentis* 1. 1. 12, *PL*, 176, 195.

70. *Ibid.* 1. 1. 19, *PL*, 176, 200.

71. *Ibid.* 1. 1. 28, *PL* 176, 204.

72. *Ibid.* 1. 11. 1–5, *PL*, 176, 343–45.

73. On this tendency in the thought of the period see Chenu, *La théologie au douzième siècle*, p. 295.

74. *Didascalicon* 1. 2; 2. 1, ed. Buttimer, pp. 6–7, 23.

75. *La théologie au douzième siècle*, p. 295.

76. *De Trinitate* 5. 1; 9.7; 10.11; *PL*, 42, 911–12, 964–65, 982–83.

77. See Chenu, *La théologie au douzième siècle*, pp. 129–35, 289–308.

78. D. E. Luscombe, *The School of Peter Abelard* (Cambridge, 1969), pp. 42–43, illustrates the diffusion of Platonist ideas in the mid-twelfth century by citing St. Bernard's use of Plato's myth of the soul's preexistence; see Bernard, *In Cantica Sermo* 27 (*PL*, 183, 915C–18B).

79. See Jerome Taylor, tr., *The Didascalicon of Hugh of St. Victor* (New York, 1961), pp. 21–22.

80. Quoted by Gregory, *Platonismo medievale*, p. 126, from Paris, B. N. lat. 8624, f. 17r.

81. On this aspect of the *De consolatione* see Györy, "Le cosmos, un songe," pp. 92–94; Peter Dronke, "L'amor che move il sole e l'altre stelle," *Studi medievali*, 6 (1965), 399–406.

82. *Cosmographia* 2. 4, *BW* p. 40.

83. Bernardus was commonly identified with Bernard of Chartres in early scholarship, an identification which was given its strongest defense by Charles-Victor Langlois, "Maitre Bernard," *Bibliothèque de l'Ecole des Chartes*, 54 (1893), 225–50, in an article still valuable for its evidence on the diffusion and attribution of Bernardus' writings. Bernardus' separate identity was established by R. L. Poole, "The Masters of the Schools of Paris and Chartres in John of Salisbury's Time," *English Historical Review*, 35 (1920), 326–31, 341–42; reprinted in Poole's *Studies in Chronology and History* (ed. A. L. Poole, Oxford, 1934), pp. 228–35, 246–47. Poole, how-

ever, manages to show only that Bernardus probably lived at Tours between 1130 and 1140, that his *Cosmographia* was presumably read to Pope Eugenius III during the latter's visit to France in 1147 or 1148, and that since the Pope did not visit Tours, Bernardus may by this time have been resident at Paris. Poole concludes by suggesting that he is possibly the Bernard who was chancellor of Chartres c. 1156, Bishop of Quimper in 1159, and whose death in 1167 is recorded in the Chartres necrology under August 4th. On this identification see also Barthélemy Hauréau, "Mémoire sur quelques chanceliers de l'église de Chartres," *Mémoires de l'Académie des Inscriptions et Belles-Lettres*, 31.2 (1884), 86–88.

For citations of Bernardus in the later Middle Ages see, in addition to the references given below, Max Manitius, *Geschichte der lateinischen Literatur des Mittelalters* (Munich, 1911–1931), III, 207.

84. For Matthew's testimony see Edmond Faral, *Les arts poétiques du xiie et du xiiie siècle* (Paris, 1923), p. 1; *idem*, "Le manuscrit 511 du 'Hunterian Museum' de Glasgow," *Studi medievali*, 9 (1936), 69–71.

Lemay, *Abu Ma'shar*, pp. 258–78, has shown that Bernardus almost certainly knew such recent works as the versions of Abu Ma'shar's *Introductorium* by Johannes Hispanus and Hermann of Carinthia (both completed during the 1130s) and probably Hermann's *De essentiis* (c. 1143); on the dating of these works see *Abu Ma'shar*, pp. 9–19. The main activity of Dominicus Gundissalinus, with whom Bernardus shows certain striking affinities, seems, however, to have been in the period after 1150; see Manuel Alonso, "Las fuentes literarias de Domingo Gundisalvo," *Al-Andalus*, 11 (1946), 172–73; "Traducciones del arcediano Domingo Gundisalvo," *Al-Andalus*, 12 (1947), 297–98.

85. See Silverstein, "Fabulous Cosmogony," 115, n. 165. The evidence is a gloss in Ms. Bodleian Laud Misc. 515, f. 188v, on the verses in *Cosmographia* 1. 3 which mention Pope Eugenius: "in whose presence this work was recited in Gaul, and gained his favor."

86. Mirella Brini-Savorelli, "Un manuale di geomanzia presentato da Bernardo Silvestre da Tours (xii secolo): l'*Experimentarius*," *Rivista critica di storia della filosofia*, 14 (1959), 285. The basis for dating the work is a passage found in certain manuscripts of the work (*ibid.*, 313–14) which makes reference to events in the Arab world during the 1160s. The validity of this evidence is questioned, however, by both C. H. Haskins, *Studies in the History of Medieval Science* (Cambridge, Mass., 1924), pp. 136–38, and Lynn Thorndike, *History of Magic and Experimental Science* (New York, 1923–1958), II, 115. Both suggest that two works of similar character have been accidentally conflated, and that there is no reason to assign so late a date to Bernardus' work.

87. *Bataille*, lines 328–30, ed. L. J. Paetow (Berkeley, 1914), p. 55.

88. See Giorgio Padoan, "Tradizione e fortuna del commento all'*Eneide* di Bernardo Silvestre," *Italia medioevale e umanistica*, 3 (1960), 227–40;

Silverstein, "Fabulous Cosmogony," pp. 97–98. That the commentary on Martianus is by the same author as that on the *Aeneid* is shown by Jeauneau, "Notes sur l'Ecole de Chartres," *Studi medievali*, 5 (1964), 845–50. My acceptance of the attribution of both to Bernardus is largely, of course, a matter of convenience, based not so much on documentary evidence as on the profound and, particularly in the case of the commentary on Martianus, the virtually unique affinity of many of their ideas with those of Bernardus. That there are real objections to this attribution is shown by Stock, *Myth and Science*, pp. 36–37, n.42.

89. See Jeauneau, "Notes sur l'École de Chartres," 846–49.

90. See Eberhard, *Laborintus*, lines 597–98, ed. Edmond Faral, *Les arts poétiques*, p. 358; Gervais, *Ars poetica*, ed. Hans Jürgen Gräbener (Munster, 1965), p. 1; also Faral, "Le manuscrit 511 du 'Hunterian Museum' de Glasgow," pp. 80–83.

91. On this development see Southern, *Medieval Humanism*, pp. 37–50, 77–83; McKeon, "Poetry and Philosophy in the Twelfth Century," pp. 297–99.

92. The moral and spiritual character of the *Cosmographia* has not always been recognized; Silverstein, "Fabulous Cosmology," pp. 92–93, summarizes the "pagan" and pantheist views of the poem which begin with the *Histoire littéraire de la France* (Paris, 1753ff.), XII, 267–72, and which are still perceptible in the assessment of E. R. Curtius, *European Literature and the Latin Middle Ages*, tr. Willard Trask (New York, 1953), pp. 108–13. The Christian character of the work was demonstrated by Etienne Gilson, "La cosmogonie de Bernard Silvestris," *Archives d'histoire doctrinale et littéraire du moyen âge*, 3 (1928), 5–24, whose reading is discussed by Silverstein, "Fabulous Cosmogony," pp. 100–12.

93. *Commentum Bernardi Silvestris super sex libros Eneidos Virgilii*, ed. W. Riedel (Greifswald, 1924), p. 1.

94. *Ibid.*, pp. 85–90.

95. See the excerpts published by Jeauneau, "Notes sur l'École de Chartres," pp. 856–64. The commentary appears in Ms. Cambridge, University Library Mm. 1.18, ff. 1–28r, and subsequent references will be to folia of this manuscript. For a summary of the commentary, see Stock, *Myth and Science*, pp. 34–35.

96. I have published excerpts outlining Bernardus' "reading" of the *De nuptiis* in an appendix to my *Platonism and Poetry in the Twelfth Century* (Princeton, 1972), pp. 271–72.

97. F. 27r, quoted and discussed in my *Platonism and Poetry*, pp. 122–24, 270–71.

98. F. lv, edited by Jeauneau, "Notes sur l'École de Chartres," p. 857, and in my *Platonism and Poetry*, pp. 124–25, 267. A similar comparison between epic poetry and spritual experience is drawn by Eriugena, *Super Ierarchiam caelestem* 2. 1 (*PL*, 122, 146): "Just as the art of the poet, by

means of feigned fables and allegorical similitudes, sets forth moral and physical knowledge as a stimulus to the human spirit (for this is the office of those 'heroic' poets who, with figural intention, praise the deeds and characters of brave men): so a sort of theological poetry, as it were, applies holy Scripture, with its fictive imagery (*fictis imaginationibus*), to the guidance of our spirit and to withdrawing it from the external bodily senses, as if from some undeveloped, childish state, into a full understanding of intelligible reality, which is, so to speak, the ripe age of the inner man."

99. F. 4r, edited and discussed in my *Platonism and Poetry*, pp. 114–15, 267–68. Cp. Bernardus' *Commentum* on the *Aeneid*, p. 55.

100. *De nuptiis Philologiae et Mercurii* 1. 7, ed. Adolph Dick (Leipzig, 1925), p. 8.

101. Ff. 15rv, edited and discussed in my *Platonism and Poetry*, pp. 115–22, 269.

102. See Jeauneau, "L'usage de la notion d'integumentum," pp. 36–53.

103. See Eriugena, *Annotationes in Marcianum*, ed. Cora E. Lutz (Cambridge, Mass., 1939), p. 13; Remigius, *Commentum in Martianum Capellam (Libri I-II)*, ed. Lutz (Leiden, 1962), p. 79; Fulgentius, *Mitologiae* 2. 2 in *Opera*, ed. Rudolph Helm (Leipzig, 1898), p. 40.

104. See Barthélemy Hauréau, ed. *Mathematicus* (Paris, 1895), pp. 7–8.

105. *Mathesis* 1. 2. 7–12, pp. 7–8.

106. *Mathematicus*, lines 639–42, ed. Haureau, p. 31. The poem is also printed among the works of Hildebert of Le Mans, *PL*, 172, 1365–80, where the passage quoted appears at c. 1377.

107. *Mathesis* 1. 4. 4, p. 12.

108. *Cosmographia* 2. 10, BW, p. 56.

109. Theodore Silverstein, ed. "Liber Hermetis Mercurii Triplicis de vi rerum principiis," *Archives d'histoire*, 30 (1955), 237.

110. *Asclepius*, esp. 4–8, in Apuleius, *De philosophia libri*, ed. Paul Thomas (Leipzig, 1908), pp. 39–43.

111. See *Asclepius* 2–3, 14, 29–35, pp. 38–39, 49–50, 67–72. Cp. *Cosmographia* 1. 4 and nn. 116–141; BW, pp. 29–32.

112. *Asclepius* 12–14, pp. 48–49.

113. *Asclepius* 10–11, pp. 46–47. See also c. 26, where the restoration of the world is prophesied, pp. 63–64.

114. *Commentarii in Somnium Scipionis* 1. 11. 11–1. 12. 2; 1. 14. 6–15, pp. 47–48, 56–58. Cp. Bernardus, *Commentum*, p. 30.

115. *Ibid.* 2. 17. 14, p. 153.

116. See *Cosmographia* 2. 4, 2. 8, 2. 10, BW, pp. 39–40, 51–52, 56.

117. On this still unedited work see Marie-Thérèse d'Alverny, "Le cosmos symbolique du xiie siècle," *Archives d'histoire*, 28 (1953), 31–81.

118. On Nature see now the valuable survey of George Economou, *The Goddess Natura in Medieval Literature* (Cambridge, Mass., 1972), whose discussion of the early history of this figure deals with her role in

poetry from Statius to Boethius (pp. 37–52) as well as with the intellectual background of the twelfth-century conception in Macrobius, Calcidius, the *Asclepius*, and Boethius (pp. 16–33).

119. *Cosmographia* 1. 4, and n. 128; BW, p. 32.124–126.

120. *Ibid*. 2. 11, BW p. 56.3–10.

121. *Ibid*. 2, 9, BW p. 53.31–32.

122. *Ibid*. 2. 14, BW p. 71.171–72.

123. *De nuptiis* 1. 8–22, pp. 9–16.

124. *Commentarius in Timaeum Platonis* 286, ed. J. H. Waszink (London-Leiden, 1962), p. 290. Cp. Aristotle, *Physics* 1. 9. 192a.

125. *Cosmographia* 1. 4, BW, p. 31.65–66. The terms *Hyle* and *Silva* are discussed by Stock, *Myth and Poetry*, pp. 97–102, who shows that, though virtually interchangeable, they can be related to specific functions, *Silva* being both more general and more clearly a part of Bernardus' moral allegory, *Hyle* being used to allude to a variety of scientific views of matter.

126. *Ibid*. 1. 4, BW, p. 30.52–54.

127. *Cosmographia* 2. 8, BW, p. 52.37–46.

128. "L'amor che move il sole e l'altre stelle," p. 414.

129. See esp. *De divisione naturae* 2.22, *PL*, 122, 566.

130. *Commentarius* 287, pp. 291–92.

131. Commentary on Boethius *De Trinitate* 2. 17, ed. N. M. Haring, *Archives d'histoire*, 31 (1956), 283. Cp. the account of the preexistence of Hyle in *Asclepius* 14, pp. 49–50.

132. *De Trinitate* 2, ed. E. K. Rand, H. F. Stewart in *Boethius. The Theological Tractates and the Consolation of Philosophy* (New York, 1918), p. 9. See Thierry's Commentary 2. 19, p. 283; Gundisalvus, *De processione mundi*, p. 23; Hermann, *De essentiis*, p. 41.

133. See especially *De processione*, pp. 39–41.

134. *De divisione naturae* 2. 15, *PL*, 122, 547; See also *De divisione* 3. 14, *PL*, 122, 663.

135. *Ibid*. 2. 16, *PL*, 122, 549.

136. *Ibid*. 2. 25, *PL*, 122, 582–83.

137. *Cosmographia* 1. 4, and n. 134; BW, p. 31.65–66.

138. On the twelfth-century tendency to regard the transforming effect of grace as a more or less inevitable stage in a "natural" process see Javelet, *Image et ressemblance*, I, 102–10; Southern, *Medieval Humanism*, pp. 48–49; Chenu, *La théologie au douzième siècle*, pp. 289–308.

139. See *Cosmographia* 1. 2, BW, pp. 9–11.

140. On this problem in the *Cosmographia* see Garin, *Studi sul platonismo medievale*, pp. 54–62.

141. See *Asclepius* 22, p. 57, and Calcidius, *Commentarius* 301, p. 303.

142. Augustine, *De Genesi ad litteram* 1. 5, *PL*, 34, 249–50; Hugh, *De Sacramentis* 1. 1. 12, *PL*, 176, 195.

143. *De divisione naturae* 2. 25, *PL*, 122, 582.

144. *Ibid.* 2. 26, *PL*, 122, 584.

145. *Cosmographia* 1. 2, BW, p. 13.138–43. Cp. the far more optimistic treatment of the same theme in *Asclepius* 16, p. 51.

146. "Fabulous Cosmogony," pp. 107–12. To the references gathered there may be added the equation of Pallas with the divine wisdom in Bernardus' commentary on Martianus. See above, n. 97.

147. Noys' association with the Platonic Archetype is clear from her "reflecting inwardly" ("introspiciens") to produce the shapes of creatures in conformity with ideal patterns, *Cosmographia* 1, 2, BW, p. 11.93. On her relation to the Second Person of the Trinity see Silverstein, "Fabulous Cosmogony," pp. 107–8. Silverstein's view is challenged by d'Alverny, "Alain de Lille et la Theologia," in *L'homme devant Dieu. Mélanges offerts au Père Henri de Lubac* (Paris, 1964), II, 121–22. She cites twelfth-century liturgical evidence of the equation of Noys, Sapientia, and Logos, but does not seem to recognize fully the fundamentally oblique character of all such quasi-theological notions in the *Cosmographia*.

148. See *Cosmographia* 1. 3, BW, p. 15.17–18, and Silverstein, "Fabulous Cosmogony," p. 108.

149. This is explicitly stated in the "Summary" which precedes the *Cosmographia*, BW, p. 5.20.

150. "La cosmogonie de Bernardus Silvestris," p. 18.

151. The term *fomes*, used by Bernardus of Endelechia (1. 2, BW, p. 14.178); of the vivifying influence of the firmament (1. 4, BW, p. 29.9); and of the sun (2. 5, BW, p. 44.135), participates in the same complex of associations that surrounds the *vitalis calor* of universal life and the *igniculus* which this kindles in the human soul. See above, pp. 40–41, and n. 159.

152. *Commentarius* 223–25, pp. 236–40.

153. *Ibid.* 225, p. 240. In *Cosmographia* 2. 2, BW, p. 35, Noys tells Natura that every creature ("res quaeque") derives "the seed and principle of its vitality" from Endelechia.

154. *Cosmographia* 1. 2, BW, p. 14.179–80.

155. *Ibid.* 2. 8, BW, p. 52.43–44.

156. See Lemay, *Abu Ma'shar*, pp. 191–93; Gregory, *Platonismo medievale*, pp. 135–37. Guillaume de Conches explicitly rejects the view of those who would identify the world soul with the sun; see Charles Jourdain, "Des commentaires inédits de Guillaume de Conches et de Nicolas Triveth sur la Consolation de la Philosophie de Boéce," *Notices et extraits des manuscrits de la Bibliothèque Impériale*, 20.2 (1862), 76.

157. See *Cosmographia* 2.5, and n. 30; BW, p. 44.140–50.

158. Eriugena, *Annotationes in Marcianum*, p. 13; Remigius, p. 13; Remigius, *Commentum*, p. 79.

159. Eriugena, glossing the phrase "Entelechiae et solis filia," explains that Entelechia is the "general" soul from which proceed, "through the

agency of the sun," the special souls proper to all parts of the world's body, "whether rational or lacking in reason." (*Annotationes*, p. 10.) *Asclepius*, 6, p. 41, explains that the spirit which vivifies all creation also bestows upon man the power of intelligence. Cicero, *De natura deorum* 2. 21. 57, calls the *ignis artificiosus* "teacher of all the other arts." In Bernardus' commentary on Martianus, f. 13r, the soul is said to derive its rational principle from Entelechia (here conceived as a type of divine wisdom) and its sensuality from the sun.

The term "igniculus" as used by Bernardus, Eriugena, and Martianus to denote the psychological effect of the *vitalis calor* recalls Cicero, *Tusculan Disputations* 3. 1, on the "parvulos igniculos" which we possess by nature but which we so obscure with bad habits that their light does not show. Here and in *De finibus* 5. 7 Cicero associates these "igniculi" with the "seeds" of virtue which reside in our *ingenium*. Cp. Quintilian, *Institutiones* 6. *Praef.* 7, on the "igenii igniculus" which in young students must be carefully nursed.

160. *Cosmographia* 2. 4, BW, pp. 39.31–40.38.

161. Such at least is the obvious implication of the "speculum Uraniae" presented to Psyche by Sophia, *De nuptiis* 1. 7, p. 8.

162. See below, Book II, n. 64.

163. This opposition reflects the twelfth-century tendency to contrast Plato the visionary with Aristotle the logician and empiricist. See Garin, *Medioevo e Rinsascimento* (Bari, 1954), pp. 57–58; Chenu, *La théologie au douzième siècle*, p. 109; McKeon, "Poetry and Philosophy in the Twelfth Century," pp. 299–301.

164. See Kristeller, "The School of Salerno," p. 515; Lemay, *Abu Ma'shar*, pp. 302–3.

165. *Cosmographia* 2. 11, BW, p. 59.87–89.

166. *Ibid.* 2. 3, BW, p. 38.91–100. One purpose of this image of the genius figure was probably to illustrate, or at least to allude to the "Chartrian" doctrine of "secondary forms," intermediary between the ideas in the divine mind and their substantial embodiments. See my article "The Function of Poetry in the *De planctu naturae* of Alain de Lille," *Traditio*, 25 (1969), pp. 112–16; Dronke, "New Approaches to the School of Chartres," p. 132. *Oyarses* stands for the Greek ουσιάρχης·

167. It is worth noting that Vulcan is assigned a function like that of Bernardus' cosmic Genius in the *Metamorphosis Goliae*, lines 37–52, ed. Thomas Wright, *Latin Poems Commonly Attributed to Walter Mapes* (London, 1841), p. 22, where he is said to have created a sumptuous palace, richly adorned and standing for the universe in both its ideal and its substantial aspects, where things appear both ideally and actually ("formam cum formatis," line 50).

168. *Cosmographia* 1. 3, BW, p. 16.53–54.

169. *Ibid.* 1. 3, BW, p. 19.133–36.

170. This is not, of course, to say that Bernardus was not seriously interested in astrology, but it may serve to set this interest in perspective. The attribution of the *Experimentarius* seems certain, and Urania in the *Cosmographia* explicitly asserts the value of astrological learning in showing "what the course of fate will grant or deny"; but it is worth noting that the knowledge offered by the *Experimentarius* is entirely practical, with virtually no relevance to philosophy or religion, and that Bernardus is at pains to stress the limits of its general usefulness (*Experimentarius*, ed. Brini-Savorelli, pp. 316–17), while acknowledging that it was able to reveal the birth of Christ. There is no reason for considering him a determinist on the basis of such evidence as his works provide.

Stock, *Myth and Science*, pp. 27–30, 129–32, distinguishes Bernardus' view from that of Hermann, who saw astrology as the culmination of cosmological knowledge, and from the determinism of Abu Ma'shar. He notes that Bernardus' intention is consistently more poetic than scientific, and points out (p. 131, n.18) that the *Mathematicus* concludes with what may be seen as an evasion of fate through the exercise of free will. In general, however, Stock grants astrology a more functional role in Bernardus' thought than my interpretation would suggest.

171. *Cosmographia* 2. 12, BW, pp. 61.55–62.
172. *Metamorphoses* 15. 75–478. With the words of Urania on mutability and death in *Cosmographia* 2. 8 cp. especially lines 165–71 of Pythagoras' discourse.
173. *Cosmographia* 2. 10, BW, pp. 55–56. Cp. *Metamorphoses* 15.838–42.
174. One may note that a link between the philosophical doctrines of the two passages is provided by Anchises' vivid discourse on the soul, *Aeneid* 6. 724–51, where the Pythagorean doctrines of purification and transmigration and the Stoic conception of the animating power of the divine intelligence are juxtaposed. Lines 724–32 of the passage were widely quoted in medieval Platonist writings; see, e.g., Eriugena, *De divisone* 1. 31, *PL*, 122, 476–77; Gregory, *Platonismo medievale*, pp. 128–29.
175. *De raptu Proserpinae* 1.76–116. On the *De raptu* and its contribution to medieval philosophical allegory see Economou, *The Goddess Natura*, pp. 46–49; Stock, *Myth and Science*, pp. 73–77.
176. *Ibid.* 3. 33–45.
177. *Ibid.* 2. 44–54; cp. *Cosmographia* 1. 3, 2. 9, BW, pp. 24. 315–25.334; 52.14–53.28.
178. *De raptu* 1. 246–72.
179. *Ibid.* 3. 18–32.
180. *Cosmographia* 2. 11, BW, pp. 58. 79–81.
181. See *Ibid.* 1. 3, BW, p. 25.333–34.
182. *De divisione naturae* 1.34, 3. 14, *PL*, 122, 479, 662–63.
183. *Ibid.* 2. 25, *PL*, 122, 582.

184. *Ibid.* 2. 18, *PL*, 122, 551. In *De divisione* 3. 20, *PL*, 122, 683, Eriugena speaks of the "gnostica theoria" by which we perceive things "in their primordial causes," a conception seemingly related to Augustine's distinction between "sermo scientiae," our perceptions of history and nature as we ordinarily know them, and "sermo sapientiae," a special vision of things as they are eternally which we arrive at through the Arts, but which is different from the mode of speculation proper to any one of the Arts. *De Trinitate* 12. 14. 23, *PL*, 42, 1010–11.

185. *De divisione naturae* 2. 25, *PL*, 122, 583.

186. *Ibid.* 1. 43, 4. 5, *PL*, 122, 485, 759–60. Cp. Bernardus' characterization of Vulcan in the commentary on Martianus, f. 15r, ed. in my *Platonism and Poetry*, pp. 116, 269: "He was granted by Jove the freedom to pursue Pallas, because he possesses from the creator a natural capacity whereby wisdom may be united with him . . . And if he does not pursue perfect knowledge by this means, he is at least capable of doing so . . . ; though he see nothing, he does not lack the capacity for vision."

187. *Ibid.* 2. 26, *PL*, 122, 584.

188. This association is of course reinforced by the fact that Noys is an angelic as well as a cosmic power, and there are a number of overlapping details in Bernardus' separate accounts of the emanation of cosmic vitality in 1. 4 and his account of the hierarchy of cosmic spirits in 2.7. On the difficult subject of Bernardus' view of "heavenly beings," see the excellent discussion of Stock, *Myth and Science*, pp. 170–78.

189. *De divisione naturae* 3. 20, *PL*, 122, 683.

190. *Cosmographia* 1. 2, BW, p. 9.3–4. See J. A. W. Bennett, *The Parlement of Foules* (Oxford, 1957), p. 108; Johan Huizinga, "Über die Verknüpfung des Poetischen mit dem Theologischen bei Alanus de Insulis," *Mededeelingen der Koninklijke Akademie van Wetenschappen, Amsterdam*, 74B, No. 6 (1932), 119–23. With such sacramental analogies, cp. Stock's suggestive remarks on the theme of "reform" in the *Cosmographia, Myth and Science*, pp. 234–37.

191. See Silverstein, "Fabulous Cosmogony," pp. 112–16.

192. On this aspect of the *De planctu naturae* see my article "The Function of Poetry in the *De planctu naturae* of Alain de Lille," *Traditio*, 25 (1969), 117–19.

193. *De planctu naturae*, ed. Thomas Wright in *Anglo-Latin Satirical Poets of the Twelfth Century* (London, 1872), II, 518; *PL*, 210, 480.

194. *Hierarchia Alani*, ed. M. T. d'Alverny, in *Alain de Lille: textes inédits* (Paris, 1965), p. 228; see also *Expositio prosae de angelis, ibid.*, p. 205, and Mlle d'Alverny's introduction, p. 95.

195. See, e.g., *Poetria nova*, lines 44–49, 60–61, 136–41, 214–18, ed. Faral, *Les arts poétiques*, pp. 200, 201, 203.

196. Bernardus was the model for many poetic exercises in the schools. See the examples quoted by Franco Munari, "Medievalia," *Philologus*, 105

(1960), 287–90; Geoffroi de Vinsauf, *Poetria nova*, 397ff., ed. Faral, *Les arts poétiques*, p. 209 (with which cp. *Cosmographia* 1. 1). On the basis of the reference in Eberhard's *Laborintus* (cited above, n. 91), Douglas Kelly suggests that Bernardus may have established the twofold classification of poetic *ornatus*; "The Scope of the Treatment of Composition in the Twelfth and Thirteenth-Century Arts of Poetry," *Speculum*, 41 (1966), 266.

197. See Alberto Varvaro, "Scuola e cultura in Francia nel xii secolo," *Studi mediolatini e volgari*, 10 (1962), 299–330; Jean Frappier, "Vues sur les conceptions courtoises dans les littératures d'oc et d'öil au xiie siècle," *Cahiers de civilization médiévale*, 2 (1959), 146–51.

198. See Reto P. Bezzola, *Les origines et la formation de la littérature courtoise en occident* (Paris, 1944–63), Vol. III, i; Varvaro, "Scuola e cultura," pp. 313–15.

199. *De amore* 1. 6, ed. E. Trojel (Copenhagen, 1892), p. 98.

200. See esp. the works of Leo Pollman cited above, n. 14.

201. *Roman de la Rose* 19115–63, ed. Felix Lecoy (Paris, 1965–70), III, 74–76; cp. *De planctu naturae*, pp. 455–56.

202. See Peter Dronke, *Medieval Latin and the Rise of European Love-Lyric* (Oxford, 1965), I, 79—87; Fernand Van Steenberghen, *Aristotle in the West* (Louvain, 1965), pp. 198–238.

203. *Medieval Humanism*, p. 43.

204. In this connection I find highly relevant to Bernardus' poem the idea of a "metaphysical pathos" to which students of the *natura rerum* are susceptible, in particular the calming and stabilizing effect of the idea of immutable order; see A O. Lovejoy, *The Great Chain of Being* (Cambridge, Mass., 1936), pp. 10–14; Hélène Tuzet, *Le cosmos et l'imagination* (Paris, 1965), pp. 10–24. Significant also is Mlle. Tuzet's distinction (pp. 216–17, 268–69) between the "felt" life of the late-classical cosmology inherited by the Chartrians, whose all-pervading *ignis* is "a sublimation of experience," and the abstract, alien character of the cosmology of thirteenth-century Aristotelianism.

## Notes to the Note on the Text

1. A number of the new readings given here for Book Two are given also by Dronke, "L'amor che move il sole e l'altre Stelle," 413, 415. See also Munari, "Medievalia," 280–83.

2. A number of emendations in punctuation are proposed by Munari, pp. 284–85.

## Notes to Dedication

1. Bernardus was not alone in his high opinion of Thierry. Hermann of Carinthia, dedicating to him his translation of the *Planisphere* of Ptolemy, declares that in Thierry the soul of Plato lives again; see Alexandre Clerval, *Les écoles de Chartres au moyen âge* (Chartres, 1895), p. 190. Clarembald of Arras, in a letter prefixed to Thierry's *De sex dierum operibus*, calls him "the foremost philosopher of all Europe;" see Häring, "Creation and Creator," pp. 183–84. See also the fine epitaph of Thierry, ed. André Vernet, in *Recueil de travaux offert à M. Clovis Brunel* (Paris, 1955), II, 660–70; Southern, *Medieval Humanism*, pp. 81–83.

2. There is perhaps a playful allusion here to Horace, *Ars poetica*, 364–65, where poems are compared to paintings, of which "one likes shadow, another likes light, and does not fear the keen scrutiny of the critic." Like some of his imitators among the composers of *artes poeticae*, Bernardus is perhaps comparing the difficulty of his task with the unruly strain in *hyle* which resists the imposition of form. The phrase "on the totality of the universe" ("de mundi universitate") in this sentence may account for the title commonly assigned to the *Cosmographia*.

## Notes to Book I: Megacosmos

1. The situation of Silva as depicted in the first two lines of this poem is strictly speaking unimaginable, as Plato had noted in dealing with it, *Timaeus* 51A. One source of Bernardus' account is probably Calcidius, *Commentarius* 222, p. 235, where the term *essentia* is said to refer in one of its senses to matter "as we consider it mentally, when unformed and still a 'silva.' In possibility it is everything that can possibly come to be, but in effect it is not yet anything, like a lump of bronze or unshaped wood." See also Hermann, *De essentiis*, pp. 37–38; Alain de Lille, "Sermo de spera intelligibili," ed. d'Alverny, *Alain de Lille: textes inédits*, p. 299. On the mode of existence of the elements in this state of *informitas* see Apuleius, *De Platone et eius dogmate* 1. 5, ed. Thomas, p. 87; Calcidius, *Commentarius* 354, pp. 344–45; Hermann, *De essentiis*, p. 37; Silverstein, "Fabulous Cosmogony," pp. 99–100. As Silverstein notes, *ibid.*, p. 99, n. 39, Bernardus' language recalls Ovid, *Metamorphoses* 1. 1–20; *Ars amatoria* 2. 467–70.

2. This motif is derived from Claudian, *De raptu Proserpinae*. See above, Introduction, Section 7.

3. On the use of the name Minerva as a key to certain of the sapiential associations of the figure of Noys see Silverstein, "Fabulous Cosmogony," pp. 110–12.

4. That Nature here demands for Silva not simply a form but a better form suggests that Bernardus has availed himself of the notion of matter as

having first been created by God "in forma confusionis." On this concept and its relation to the supposed dualism of Bernardus' thought, see Silverstein, "Fabulous Cosmogony," pp. 99–103.

5. This difficult sentence seems to point to the existence of a basic aspiration in Nature which she herself does not understand, the "sacred and blessed instincts" commended by Noys in her reply to Nature's prayer. Cp. the final sentence of Boethius, *De Trinitate*, trans. Steward-Rand, p. 31: "But if human nature has failed to reach beyond its limits, whatever is lost through my infirmity must be made good by my intention."

6. See Vergil, *Aeneid* 6. 731, and Silk, p. 177, where God is described as "not refusing to shape his creation to His own image and likeness, so far as the material itself is capable of undergoing it . . ."

7. Cp. *Timaeus* 29D–30A.

8. *discolor usiae vultus*. The term "usia" here seems to me to recall the opening book of the *De divisione naturae* of Eriugena, esp. cc. 3, 34, 48–63, *PL*, 122, 443, 479, 490–508, where οὐσία is defined as an essence preexistent to the *coitus* of qualities which produces matter. In connection with this phrase Gilson, "La Cosmogonie de Bernardus Silvestris," p. 10, cites Calcidius, *Commentarius* 289, pp. 293–94, which alludes to the Stoic identification of Silva and *essentia*, but Eriugena's conception is truer to Bernardus' presentation of the evolution of life. See also Eriugena's allegorical reading of the opening of *Genesis*, as describing the emergence of the primordial causes from the abyss of the divine wisdom, *De divisione naturae* 2. 17, *PL*, 122, 550.

9. On this "longing" in Silva see Calcidius, *Commentarius* 286–87, pp. 290–92.

10. Cp. *De divisione naturae* 2. 16, *PL*, 122, 549, quoted above, p. 37.

11. On the eternity of matter see below, nn. 133–34.

12. The image suggests the *informitas* and at the same time the innate vitality of a universe whose "speech" has not yet been made articulate by the impress of the *Logos*. See also Calcidius, *Commentarius* 297, pp. 299–300.

13. The elements are classified in terms of the Aristotelian Four Causes by Thierry, *De sex dierum operibus*, pp. 195–96.

14. On the spontaneity of the elements cp. *Timaeus* 53A, Guillaume de Conches, *Glosae super Platonem* 176, p. 290.

15. On the opposition of privation (*carentia*) to form (*species*) in relation to Silva, see Calcidius, *Commentarius* 286–88, pp. 290–93, commenting on Aristotle, *Physics* 1. 9; Gundisalvus, *De processione mundi*, pp. 25–27.

16. *vultu blandiore*. Other translations of these words are possible, but it seems appropriate to the boldness of Bernardus' conception that his Noys should have been brooding over the vast solitude of the precreation until aroused by the energy and enthusiasm of Nature. Nature had noted that she seemed "old and sad" ("vetus et gravis," 1.1; BW, p. 8.57).

17. The image is used of *ratio* by Calcidius, *Commentarius* 104, p. 153.

18. *uteri mei beata fecunditas.* On this "Marian" conception of Nature see the references gathered in Introduction, n. 192, above; also Rudolf Krayer, *Frauenlob und die Natur-Allegorese* (Heidelberg, 1960).

19. Cp. Cicero's characterization of Nature as "concerned to provide all that is useful or serviceable," *De natura deorum* 2. 21. 58.

20. *Dei ratio profundius exquisita. Ratio* as used of Noys implies "relation," "method," or "plan," and suggests that her role is that of a particular function of God's wisdom, and not that wisdom conceived as the Second Person of the Trinity. On *ratio* see Waszink, ed. Calcidius, *Commentarius*, pp. 420–21; *De sex rerum principiis* 14, ed. Silverstein, p. 248, where *ratio* is defined as "a force proceeding from a cause, ordering all things from the beginning;" Javelet, *Image et ressemblance*, I, 169–76.

21. See Calcidius, *Commentarius* 276, p. 281, who quotes *Proverbs* 8. 22–25, and remarks that the verses clearly show the wisdom of God to have been established as the *primordium* of the universe: "from which," he says, "it appears that this wisdom was in some sense created by God, but not at any point in time. For there can have been no time when God was without wisdom." Bernardus recalls this passage also in glossing the Entelechia of Martianus' *De nuptiis;* Ms. Cambridge U. L. Mm. 1.18, f. 13r, ed. in my *Platonism and Poetry*, p. 268: "Wisdom [may be interpreted as] 'the age of God' or 'eviternity' ('Dei etas quasi evitas') in that it exists apart from any beginning or end or movement of time. For if it had had a beginning or were ever to have an end, God would not be always (i.e., eternally) wise."

22. With this account of Noys' activity cp. Calcidius, *Commentarius* 176, pp. 204–5, where the names of Noys and Providence are associated and where Providence is characterized as transmitting to the universe an integrity derived from the goodness of God; also *Asclepius* 26, p. 64, where the relation of divine deliberation and will to created life is discussed. See also below, 1. 4, and n. 125.

23. Bernardus would here seem to be steering a middle course between the Stoic and Pythagorean views of Silva as reported by Calcidius, *Commentarius* 296–97, pp. 298–99, the Stoics viewing her as neither good nor bad, Pythagoras seeing her as actively malign.

24. The image comes from Calcidius' paraphrase of Aristotle, *Physics* 2. 8. 199a, in *Commentarius* 287, p. 292.

25. See the opening of 2. 13, BW p. 61.1–10.

26. The language is borrowed from Calcidius, *Commentarius* 302–3, 316–17, pp. 304–5, 313–14. Cp. Gundisalvus, *De processione mundi*, p. 31.

27. Cp. Calcidius, *Commentarius* 274, p. 278.

28. Cp. *Timaeus* 43A, Calcidius, *Commentarius* 204, p. 223.

29. Cp. Calcidius, *Commentarius* 278, 292, pp. 282–83, 295. On the "seal" ("signaculum") of form see below, n. 35.

30. Calcidius, *Commentarius* 282, p. 285, reports Empedocles' use of this image.

31. The sentence echoes the close of *Asclepius* 15, p. 50. Cp. also Calcidius, *Commentarius* 296–97, pp. 298–99.

32. With the rest of this paragraph cp. Macrobius, *In Somn. Scip.* 1. 6. 23–28, pp. 22–23.

33. Cp. *Timaeus* 31B–32C.

34. Cp. *Timaeus* 30CD. The term *notio*, which I have translated as "idea" in this sentence, is one of a number of terms drawn from Eriugena and the pseudo-Dionysius which assume highly technical meanings in the "Scotist" ambiance of mid-twelfth-century theology. (For *notio* see, e.g., *De divisione naturae* 4. 7, 8, 13, *PL*, 122, 768B, 774A, 802C; *De praedestinatione* 2. 4, *PL*, 122, 362D.) Javelet, *Image et ressemblance*, 2, 71–72 (ch. 3, n. 9), quotes the distinction of Simon of Tournai, between the Platonic conception of *notiones* "through which the future situation of the world and of worldly things became known to God," and the truer sense in which these "thoughts" are merely a way of speaking about the Persons of the Trinity. Bernardus may have deliberately chosen an ambiguous term to characterize the preconceptions of the divine mind.

35. *signaculis idearum*. On the term *signaculum* see Gilson, "La cosmogonie de Bernardus Silvestris," p. 15, n. 1, who cites Calcidius, *Commentarius* 337, pp. 330–31 as a source for the doctrine presented. On its relation to the Chartrian doctrine of "secondary forms" or "images" intermediary between the divine ideas and material life see Silverstein, "Fabulous Cosmogony," p. 113, who notes the use of the term *signaculum* in Calcidius, *Commentarius* 327–28, pp. 322–23, in connection with *Timaeus* 49A–50C. The same Chartrian conception is perhaps reflected also in the work of the celestial Genius of *Cosmographia* 2. 3; see my article, "The Function of Poetry in the *De planctu naturae*," pp. 112–17; Peter Dronke, "New Approaches to the School of Chartres," *Anuario de estudios medievales*, 6 (1969), 131–32; and above, Introduction, n. 166.

For patristic uses of *signaculum* see Albert Blaise, *Dictionnaire latin-français des auteurs chrétiens* (2d ed., Paris, 1967), s.v. Like *notio*, it occurs also in Eriugena (*De divisione* 4. 13, *PL*, 122, 801–2), and becomes a technical term in twelfth-century theology; see Javelet, *Image et ressemblance*, I, 162–64. But as used by Bernardus it seems only to allude obliquely to a theological context. Cp. the various uses of the term by Alain de Lille, discussed in my "The Function of Poetry in the *De planctu naturae*," p. 104, n. 76.

36. With this paragraph cp. Calcidius, *Commentarius* 22, pp. 72–73; *Timaeus* 31B–32C; Macrobius, *In Somn. Scip.* 1. 6. 23–28.

37. With the rest of this paragraph cp. Hermann, *De essentiis*, pp. 43–44, and Bernardus' account of the work of Physis in 2. 13.

38. This phrase is borrowed from *Asclepius* 3, p. 39, and also of course suggests *Genesis* 1.

39. Cp. *Timaeus* 33AB; Calcidius,.*Commentarius* 24, p. 75.

40. Cp. Calcidius, *Commentarius* 181, 192, pp. 209, 215.

41. Cp. Calcidius, *Commentarius* 199, pp. 219–20.

42. This phrase and the characterization of Endelechia as "illumination" in the next paragraph suggest the neo-Platonism of the pseudo-Dionysius and Eriugena; cp. *De divisione naturae* 1. 75, 2. 22, *PL*, 122, 520–22, 565; also Eriugena's translation of *Celestial Hierarchy* 13, 15, *PL*, 122, 1061–62, 1065–66. With the fountain image cp. *Asclepius* 3, p. 38.

43. A number of phrases in this paragraph occur also in *Dex sex rerum principiis* 29, 32–34, pp. 249–50. Professor Silverstein has kindly informed me that he now thinks it probable that the anonymous author of this treatise was the borrower, rather than Bernardus.

44. This phrase is used of the divine goodness in *Asclepius* 20, p. 56.

45. *rerum cognitio praefinita*. On the term *praefinita* see Gregory, *Anima mundi*, pp. 72–75. Thierry, *De sex dierum operibus*, p. 196, speaks of God's "aeterna praefinitio" regarding all things.

46. Cp. the accounts of the "forma divinitatis," in which all forms of life are reflected as if in a mirror, in Hermann, *De essentiis*, p. 42, Gundisalvus, *De processione mundi*, p. 40.

47. Cp. *Asclepius* 39, p. 79.

48. *quadam emanatione*. Cp. *Wisdom* 7. 25.

49. *fomes vivificus*. On the implications of such language see above, Introduction, n. 151.

50. Cp. *Timaeus* 32C. On the "marriage" of form and matter cp. Calcidius, *Commentarius* 301–2, pp. 303–4; Gundissalinus, *De processione mundi*, p. 39; Hermann, *De essentiis*, pp. 43–44.

51. Cp. *Romans* 8. 22.

52. Cp. *Timaeus* 43C.

53. Does this use of *tabernaculum* hint at man's future violation of the "laws of hospitality"? Cp. the allusion to man as "guest" in Bernardus' account of Paradise, 1. 3, above, p. 83, and below, Book Two, n. 73. In the *De planctu naturae* of Alain de Lille Nature herself becomes a shrine or temple, where Genius, as priest, pronounces a decree of excommunication against sinful man.

54. With this and the following paragraph cp. Macrobius, *In Somn. Scip.* 1. 14. 4–15, pp. 56–58.

55. Following upon Bernardus' emanationist account of the creation of Endelechia, this reference to the "explication" of creation in space and time is strongly recollective of the similar language of Thierry's commentary on Boethius *De Trinitate* 2. 9, 20–21, pp. 280, 284, where God is said to "enfold" the totality of created life of which creation is the "unfolding." This unfolding or "necessary complication" is variously called, says Thierry, by

such names as "natural law," "nature," "the world soul," *"heimarmene,"* "fate," and "divine understanding," which suggests the hierarchical participation of all these powers in the creative process. The *locus classicus* is, of course, *De consolatione* 4, pr. 6.

56. This reference, apparently to the upper heavens and perhaps the firmament as fiery in composition, is not consistent with Bernardus' account of the quintessence in 2. 3. See below, Book Two, n. 8.

57. Bernardus is apparently substituting the angelic hierarchy for the "gods" of *Timaeus* 39E—40A.

58. Cp. Honorius, *De imagine mundi* 1. 139–40, *PL*, 172, 146, who mentions the "spiritual heaven" inhabited by the nine hierarchies and then the "heaven of heavens" set far beyond, where the king of the angels resides. In Honorius' *Elucidarium* 1. 3, *PL*, 172, 1111, God is said to inhabit the "intellectual heaven."

59. Silverstein, "Fabulous Cosmogony," p. 108, cites these lines as evidence for the disassociation of Noys from the Trinity. Noys' angelic role is further suggested by the incorporation of certain phrases from Calcidius' account of her activity into Bernardus' account of the highest rank of spirits in 2. 7. See also below, Book Two, n. 45.

It must be noted, however, that Noys' presence in the heaven of the Thrones does not preclude her identification with the divine wisdom in itself. For Eriugena, the heaven of the Thrones is where God Himself sits in judgment on creation; see the epigram prefatory to his translation of the *Celestial Hierarchy*, *PL*, 122, 1038, and his commentary on c. 7 of this work, *PL*, 122, 1050.

60. *Cum propriae causas utilitatis habent.* The point seems to be that the Virtues illustrate the intrinsic power of virtue in general.

61. The unusual syntax of this couplet is evidently modeled on Statius, *Thebaid* 1. 81; see Franco Munari, "Medievalia," pp. 280–81.

62. On the function of these lines see above, Introduction, Section 7. Though Calcidius, *Commentarius* 126, pp. 169–70, comments on the star of Bethlehem, an equally likely source for Bernardus' allusion here is the *Introduction maius* of Abu Ma'shar (Albumazar) 6. 2 (ed. Venice, 1506, f. e4v), perhaps by way of Hermann, *De essentiis*, p. 29. See Silverstein, "Fabulous Cosmogony," p. 96, n. 27.

63. This couplet was presumably added to the poem for the occasion of its public recitation before Pope Eugene III; see above, Introduction, n. 85. The lines are almost identical with a couplet on the hypothetical perfection of Patricida in the *Mathematicus*, lines 505–6, ed. Hauréau, p. 28; *PL*, 171, 137, and the association is perhaps intended to reinforce the lavish flattery of the lines, casting Eugene as the new hero made possible by Christ.

64. *et aeternum volvere stare fuit.* The syntax mimes the idea of circular motion as the image of eternity; see *Timaeus* 37D–38C.

65. In this description of the colural arcs, and in 2. 1, Bernardus follows

Macrobius, *In Somn. Scip.* 1. 15. 14, p. 63. In 2. 3, however, he presents the view put forward also by Guillaume de Conches, *De philosophia mundi* 2. 14, *PL*, 172, 61, that the arcs of the colures do in fact attain completion.

66. The rather vague reference to "the position of Egypt" echoes Hyginus, *Astronomicon* 2. 19.

67. By virtue of his location in relation to Ara, this must be Sagittarius, mentioned again below, rather than Centaurus.

68. This is the "Southern" Fish, which Vergil calls "piscis aquosus," *Georgic* 4. 234. Cp. Hyginus, *Astronomicon* 2. 41.

69. On this paragraph see above, Introduction, Section 7.

70. On the seven climates see Isidore, *Etymologiae* 3. 42. 4.

71. "Terebinthus olens" is simply listed with the other names in this passage, suggesting that Bernardus understood it to be the name of a mountain. He may perhaps have misread Pliny, *Naturalis historia* 16. 18. 73, where it is said that "the terebinth loves the mountains."

72. Syntactically this clause could refer to the Alps, but I am persuaded by the glossator of Ms. Oxford, Bodleian Laud Misc. 515, who understands it as describing the Pyrenees.

73. The sequence of details in this paragraph, and the unusual use of the term *articulus* in connection with mountains, suggest a deliberate contrast with the passage in praise of Italy at the close of Pliny's *Naturalis Historia* 37. 201, which refers in succession to "her abundance of waters, the healthy air of her groves, the valleys which link her mountains, the harmlessness of her wild beasts, the fertility of her soil, the lushness of her meadows."

74. The use of the verb "extruitur" suggests that tusks are the "ossa" referred to here, but cp. also *Job* 40. 18.

75. Isidore, *Etymologiae* 12. 1. 39, reports that male *onagri*, each of whom presides over a herd of females, are wont to eat the testicles of newborn males, so that mothers, fearing this fate for their offspring, hide them away in deserted places.

76. Pliny, *Nat. Hist.* 8. 38. 137, reports that the urine of the lynx crystallizes into a precious gem; having learned that this gem is coveted by man, he is very secretive about where he urinates. Cp. Isidore, *Etymologiae* 12. 2. 20. To accept this account as the basis of Bernardus' description is regrettably to reject C. S. Lewis' charming reading of the couplet, *Allegory of Love* (Oxford, 1936), p. 93.

77. The beaver's voluntary self-castration is reported by Isidore, *Etymologiae* 12. 2. 21.

78. There is, so far as I know, no difference between the beaver (*beber*) of this line and the beaver (*castor*) mentioned just above.

79. A slightly garbled version of this couplet and the following line is quoted by Petrus Cantor, *Verbum abbreviatum* 85, *PL*, 205, 255, as an illustration of excessive "sumptuosity" in dress.

80. For the Latin of this couplet, not given in the Barach-Wrobel text, see the textual notes, above, p. 64.

81. Semiramis. See Ovid, *Metamorphoses* 4. 57–58.

82. See Lucan, *De bello civili* 8. 595–608.

83. The sources of this couplet are obscure. Munari, "Medievalia," p. 283, cites Pliny, *Nat. Hist.* 5. 18. 74, where, however, the river is not mentioned by name. In *II Kings* 5. 12, the leper Naaman resists Elisha's command to wash himself in the Jordan, arguing that "Abana and Pharpar, rivers of Damascus," are clean enough. Since Shiloah is mentioned in the next couplet, Bernardus may also be recalling *Isaiah* 8. 1–8.

84. St. Maurice and his "Theban Legion."

85. See Ovid, *Metamorphoses* 2. 235–324.

86. Pliny, *Nat. Hist.* 13. 18. 109.

87. See Vergil, *Georgic* 4. 47.

88. The source of this threefold classification is Vergil, *Georgic* 2. 9–21.

89. This prosaic dismissal of the pear may be a little pedantic joke on the part of a professional rhetorician who would of course know the many varieties of pear and their elaborate names. See Pliny, *Nat. Hist.* 15. 15. 53; Jean de Hautville, *Architrenius* 4. 46–52, ed. Thomas Wright, *Anglo-Latin Satirical Poets*, I, 294.

90. Cp. Pliny, *Nat. Hist.* 23. 73. 141.

91. The reference is perhaps to the legend of Phyllis' metamorphosis into an almond tree; see Hyginus, *Fabulae* 59, 243; Servius, *ad Eclogam* 5. 10. But since the almond is mentioned a few lines below, the allusion may also possibly represent an imperfect recollection of Vergil, *Eclogae* 2. 52, where chestnuts are said to be the favorite food of Amaryllis.

92. Cp. Pliny, *Nat. Hist.* 12. 14. 29.

93. *fontanis marcida guttis*. Perhaps a metonymy, alluding to the association of this grove with human sacrifice.

94. Cp. Pliny, *Nat. Hist.* 19. 38. 126.

95. Cp. Macer Floridus, *De viribus herbarum*, ed. Louis Choulant (Leipzig, 1832), line 882.

96. See *ibid.* 1033–34.

97. Cp. Isidore, *Etymologiae* 17. 9. 43.

98. Cp. Macer Floridus, *De viribus* 7–8.

99. Cp. Pliny, *Nat. Hist.* 8. 27. 99; Isidore, *Etymologiae* 12. 4. 46–47.

100. Cp. Pliny, *Nat. Hist.* 20. 22. 245.

101. Cp. *ibid.* 25. 8. 81; Isidore, *Etymologiae* 17. 9. 49.

102. Cp. Macer Floridus, *De viribus* 200–1.

103. Cp. *ibid.* 331–35.

104. Cp. *ibid.* 409–10.

105. Cp. Pliny, *Nat. Hist.* 26. 8. 62–71.

106. Cp. *ibid.* 18. 35. 361; Isidore, *Etymologiae* 12. 6. 11.

107. *morius insipiens.* Bernardus is perhaps playing on the similarity of "morius" to "morio," "fool," and of "insipiens" to "insipidus," "tasteless."

108. Cp. Pliny, *Nat. Hist.* 2. 41. 109; Isidore, *Etymologiae* 12. 6. 47–48.

109. The attribution of aphrodisiac qualities to the stickleback is perhaps due to the fact that its name, *stincus,* is also a name for the herb satiricon mentioned above.

110. The association of the carp (*darsus,* OF *darset*) with the Loire occurs again in the declaration, apparently proverbial, that "a woman is more elusive than the carp in the Loire," quoted from a medieval manuscript by F. E. Godefroy, *Dictionnaire de l'ancienne langue française* to illustrate *darset.*

111. For the Latin of this couplet, not given in Barach-Wrobel, see the textual notes, above, p. 64.

112. See Lucan, *De bello civili* 5. 716; Isidore, *Etymologiae* 12. 7. 14.

113. Cp. Pliny, *Nat. Hist.* 10. 41. 120.

114. Cp. *Job* 38. 41.

115. The doctrine stated in this paragraph, that the movement of the firmament and the stars governs the processes of elemental life in the world, receives perhaps its clearest formulation in Abu Ma'shar, *Introductorium* 1. 2, f. a5r: "For it is the perpetual circling of the upper world about the lower, since its motion draws that world into subjection to it, that agitates the materials of universal life and effects the mixture of those active and passive qualities which are the causes of all generation." See Lemay, *Abu Ma'shar,* pp. 257–71. See also Bernardus' *Experimentarius,* ed. Brini-Savorelli, pp. 312–13; Hermann, *De essentiis,* pp. 85–86; Gundisalvus, *De processione mundi,* pp. 54–55; *De vi rerum principiis* 71, p. 254.

116. *Tamquam ex deo vitae.* Cp. *Asclepius* 3, p. 38: "The firmament, the god of the sensible universe ("sensibilis deus"), assigns life to all bodies whose growth and diminution are in the charge of the sun and moon." *Asclepius* 19, p. 54, speaks of the planetary usiarchs as "sensible gods," the chief of whom is Jupiter, who "bestows life upon all creatures through the firmament." Note that the first of these passages is echoed also in Bernardus' comparison of Endelechia to a flowing fountain (see above, n. 42), and so constitutes a poetic link among the various cosmological doctrines presented in the *Megacosmos.*

117. With this and the following sentence cp. Firmicus Maternus, *Mathesis* 1. 5. 10–11, p. 17.

118. The most obvious source of these lines is *Asclepius* 2, p. 38, where the pervasive influence of fire on the inferior elements is discussed, but the idea is widespread in the early twelfth century. Cp. the very similar account in Abu Ma'shar, *Introductorium* 1. 2, f. a5v, quoted and discussed by Lemay, *Abu Ma'shar,* p. 257, and the description of the roles of active and passive natures in creation in Hermann, *De essentiis,* pp. 63–64; also Thier-

ry's discussion of the roles of active fire and passive earth, *De sex dierum operibus*, pp. 192–93. On the possible influence of Abu Ma'shar's conception on those of Thierry and others, see Lemay, *Abu Ma'shar*, pp. 181–82. On the Stoic basis of this conception, see Stock, *Myth and Science*, pp. 138–41.

119. Cp. *De sex rerum principiis* 71, p. 254.

120. Cp. Eriugena, *De divisione naturae* 1. 30, *PL*, 122, 476.

121. Cp. *Timaeus* 37C–38C.

122. This sentence is a patchwork from several sources. The conception of God as the primary substance (*usia*) is presumably derived from the opening book of Eriugena's *De divisione naturae;* "eternal permanence" echoes *Asclepius* 41, p. 81; "simplicity fecund of plurality," while perhaps ultimately Scotist, is also strikingly reminiscent of Thierry's account of the generative activity of the divine mind in his commentary on Boethius *De Trinitate* 2. 32–33, p. 288, and nn. 4–7; the remainder of the main clause echoes *Asclepius* 14, p. 50.

123. Cp. Seneca, *Quaestiones naturales* 2. 45.

124. The qualification calls attention to itself, suggesting that Bernardus intends by it to point up the distinction between Sapientia as manifested in Noys and the true Logos.

125. This account of the divine act recalls Noys' description of her activity in 1. 2, BW, p. 9. 8–15. Bernardus' language echoes *Asclepius* 26, p. 64.

126. An echo of *Asclepius* 20, p. 56.

127. Very similar language is used of the action of the Trinity in Bernardus' commentary on Martianus, f. 27r; see my *Platonism and Poetry*, pp. 123, 271.

128. The terms which I translate as "elementing" and "elemented" (*elementans, elementata*) seem to have been coined by Guillaume de Conches in his *De philosophia mundi*. See Dronke, "New Approaches to the School of Chartres," pp. 128–32, who argues convincingly against Lemay's attribution of the term to Johannes Hispalensis in his translation of the *Introductorium maius* of Abu Ma'shar; see Lemay, *Abu Ma'shar*, pp. 25, 74–75. On *elementata* see also Silverstein, "Elementatum: Its Appearance among the Twelfth-Century Cosmogonists," *Medieval Studies*, 16 (1954), 156–62. The evolution from precorporeal matter to created life is traced in comparable but more concrete terms by Hermann, *De essentiis*, pp. 39–40; see also Gundisalvus, *De processione mundi*, p. 54, and his *De anima*, 7, ed. J. T. Muckle, *Medieval Studies*, 2 (1939), 53; Guillaume de Conches, *Glosae super Platonem*, pp. 268–69.

On a less technical level, the language echoes the close of *Asclepius* 3, p. 39: "Nature, imaging (*imaginans*) the world with species through the four elements, brings forth all things below the heavens that they may be found pleasing in the sight of God." The texture of the words "elementans," "elementa," "elementata," gives the effect of emanation and suggests an imaginative variation on the fourfold hierarchy of *naturae* in the *De divisione* of Eriugena. On another such play see below, Book Two, n. 38.

129. Lemay, *Abu Ma'shar*, p. 265, has detected in these lines an implicit criticism of the "medici" of the day, whose interest in the principles of physical nature seemed to be supported by an inadequate understanding of the celestial source of these principles. Lemay compares Hermann, *De essentiis*, p. 52. See also Introduction, n. 25. With the general sense of this sentence and the rest of the paragraph cp. *De vi rerum principiis* 39–40, p. 250.

130. The special role of the sun and moon is asserted by Abu Ma'shar, *Introductorium* 3. 4, f clr; see also Hermann, *De essentiis*, p. 52, and Lemay, *Abu Ma'shar*, pp. 253–54.

131. This definition of Nature is close to that of the nature who endows the elements with the power to express their qualities in *De sex rerum principiis* 24–26, p. 249.

132. Cp. Firmicus, *Mathesis* 1. 5. 10, p. 17. As the source of Bernardus' "universal bonds" ("ligamina universa") Lemay, *Abu Ma'shar*, pp. 267–68, suggests the phrase "sibi alligatum trahat" which I have translated as "draws . . . into subjection to it" in the passage quoted in n. 115 above.

133. The ideas presented in this and the previous sentence are difficult to trace to a precise source. Their general purpose seems to be to show how all the *principia* of created existence participate in a common life. In this the passage recalls Eriugena's account of the divine preexistence of the "primordial causes" of all corporeal and incorporeal life, including primordial matter, in the Word, *De divisione naturae* 2. 22, PL, 122, 566; also Silk, pp. 157–58, 177; at p. 156 of this work the *vita* of *John* 1. 3–4 is glossed as representing the idea of the world in the mind of God.

134. Cp. *Asclepius* 14, pp. 49–50, where the universe is traced to the original existence of God, ὕλη (Hyle), and a spirit which is the companion of, or instilled in Hyle. This Hyle or "Mundi natura" and the spirit which exists with or within it, "although they cannot be said to owe their origin to 'birth,' yet possess within themselves the power and the nature of birth and procreation."

135. Cp. Silk, p. 157, enumerating the works of God (the divine mind, the universe, the world soul, the chaos or ὕλη) and those of nature ("when anything takes rise from the sowing of seed or from some power of its own"). The formulation is probably based on Calcidius, *Commentarius* 23, p. 74. With the paragraph as a whole cp. Macrobius, *In Somn. Scip.* 1. 14.8–16.

136. With this and the next two sentences cp. Firmicus, *Mathesis* 1. 5. 7–9, p. 16; Cicero, *De natura deorum* 2. 16; see also Lemay, *Abu Ma'shar*, pp. 191–93.

137. Cp. *Asclepius* 40, p. 80.

138. Cp. *Timaeus* 32CD; Calcidius, *Commentarius* 24, p. 75.

139. This and the following sentence occur virtually in this form in *De sex rerum principiis* 35, 36, p. 250. Cp. Silk, pp. 157–58.

140. Cp. Calcidius, *Commentarius* 105, p. 154; *Timaeus* 37C–38C. Silverstein, "Fabulous Cosmogony," p. 103, detects in this passage an allusion to Eriugena's distinction (*De divisione naturae* 2.21, 23; 3.3, 5, etc.) between the absolute eternity of God and the "not wholly coeternal" eternity of the divine exemplars of created life. Dronke, "New Approaches to the School of Chartres," pp. 131–32, n.49, points to the similar discrimination of *coeternitas, eternitas,* and *concreatio* attributed to Bernard of Chartres by John of Salisbury (*Metalogicon* 4.35), and asserts that this view "provided a philosophical basis for the whole poetic mythology of the creative process elaborated by Bernardus Silvestris and Alan of Lille," by defining a distinct realm of activity for Nature, Genius, etc., on the level of *eternitas,* ideal existence subsisting "in the secret depths of the divine mind."

141. Though the rhythmical elaboration of this paragraph of the relation between the sensible universe and its exemplar is Bernardus' own, certain hints may have been borrowed from the discussion of the relations between the temporal and the eternal in *Asclepius* 29–35, pp. 68–75. See, e.g., *Asclepius* 29, p. 68: "If then the universe has ever been, is, and ever will be a living animal, nothing in the universe is mortal . . . Therefore the universe must be wholly filled with life and eternity, if it is necessarily to live eternally"; also *Asclepius* 30, p. 68: "for the universe is generated in the very vitality (*vivacitas*) of eternity, and in this same vital eternity it resides." *Asclepius* 32, p. 71, describes eternity, "wholly filled with [the ideas of] all sensible things and all knowledge (*disciplina*)," and 33, p. 72, shows how the world emulates this fullness: "for in order that the universe itself may be full and perfect all the parts of the universe are wholly filled with bodies differing in form and quality, possessing their own shapes and sizes."

142. This sentence occurs also as *De sex rerum principiis,* 37, p. 250. Cp. Silk, pp. 175–76, on the emergence of time from, and its resolution in the *aevum.*

143. The discussion of time and eternity in this paragraph is based on, and is largely a patchwork of phrases from *Asclepius* 30–31, pp. 68–70.

144. This sentence occurs as *De sex rerum principiis* 38, p. 250.

145. Cp. Cicero's analysis of the relations of Caelus, Saturn, and Jove, *De natura deorum* 2. 24. 64: Saturn (i.e., time) castrates Caelus (celestial fire) so that this power will be forced to cooperate with others in procreation. Jove represents celestial order which in its turn regulates the activity of Saturn.

146. The phrase comes from *Asclepius* 20, p. 56.

147. The passage recalls Boethius' account of the relations of Providence and Fate, *De consolatione philosophiae* 4, pr. 6, and Macrobius' account of

the "golden chain" which emanates from the divine mind, *In Somn. Scip.* 1. 14. 15, p. 58. The process here traced is summarized under the general heading of εἱμαρμένη or "necessity" in *Asclepius* 39, p. 79. Cp. Thierry's commentary on Boethius *De Trinitate* 2. 21, p. 284, defining that necessity "which some call 'natural law,' some 'nature,' some 'the world soul,' some 'natural justice,' some 'heimarmene'; but others call it 'fate,' others 'the Parcae,' others 'God's understanding.'" See also *Timaeus* 41D and Gilson, "La cosmogonie de Bernardus Silvestris," p. 18.

148. On Nature as artisan (*artifex*) see Silverstein, "Fabulous Cosmogony," pp. 104, 106, and nn. 82, 98; also the instances cited from Pliny's *Naturalis Historia* by Gilson, "La cosmogonie de Bernardus Silvestris," p. 22. It is this aspect of Nature which is represented by the Physis of the *Microcosmos*.

149. The definition of "imarmene" which perhaps best fits Bernardus' scheme is that of Cicero, *De divinatione* 1. 55. 125. See also pseudo-Aristotle, *De mundo* 38, in Apuleius, *De philosophia libri*, p. 174; Gilson, "La cosmogonie de Bernardus Silvestris," p. 18; and the summarial accounts of Thierry and the *Asclepius* cited above, n. 147.

150. Imarmene's "joining and rejoining" here is very close to the account of the universe as perpetually "joined and rejoined" in Silk, p. 158.

## Notes to Book II: Microcosmos

1. On man as the completion of the universe see *Asclepius* 10, pp. 45–46. On the implications of this idea for twelfth-century thought see the important discussion of Chenu, *La théologie*, pp. 52–61.

2. Gregory, *Platonismo medievale*, p. 95, quotes from an anonymous commentary on the *Timaeus* contained in Ms. Vienna, Nationalbibliothek 278, which explains (f. 47v) that Plato's Demiurge entrusted the creation of subcelestial animals to lesser powers because "it was not right that that which was eternal and without beginning should create things mortal and temporal," but later adds (f. 57r) that the *opifex* reserved to himself the creation of the immortal part, "the soul and the permanent virtues of the soul, reason and understanding."

3. The situation recalls that of Mercury and Virtue seeking the abode of Apollo in the opening book of the *De nuptiis*.

4. On *Anastros* ("without stars") see Martianus, *De nuptiis* 8. 814, p. 431.

5. See above, Book One, n. 64.

6. The basis of this account is Macrobius, *In Somn. Scip.* 1. 11. 11–1. 12. 2, pp. 47–48, also drawn on by Bernardus in the *Commentum* on the *Aeneid*, p. 30, though probably by way of Guillaume de Conches; see Jeau-

neau, "L'usage de la notion d'integumentum," pp. 42–43. For Eriugena, *De divisione naturae* 2. 25, *PL*, 122, 582–83, man's present fleshly body is imposed upon him as a punishment for sin. Cp. the anonymous *De mundi constitutione*, *PL*, 90, 901, and Thierry's inclusion of "inferno" with "chaos" and "silva" among the names of primordial matter, in his commentary on the *De Trinitate* 2. 18, p. 283. On the theme in general see Courcelle, "Tradition platonicienne et traditions chrétiennes du corps-prison," *Revue des études latines*, 43 (1965), 406–43.

7. On the term see Macrobius, *In Somn. Scip.* 1. 11. 6, p. 46.

8. Bernardus' account of the quintessence, based verbally on pseudo-Aristotle, *De mundo* 1, p. 138, does not seem to reflect a consistent position. In *Cosmographia* 1. 3, above, p. 115, BW, p. 15. 5, he seems to suggest that the firmament is fire, "essentia purior ignis"; see also the reference to the stars as "eternal fire" in the passage quoted above, Introduction, p. 47. The doctrine of the quintessential composition of the stars and planets appears in Abu Ma'shar, *Introductorium* 1. 2, f. a5r, and is opposed by Guillaume de Conches in his *Dragmaticon*, cited by Lemay, *Abu Ma'shar*, p. 262, on the grounds that non-elemental bodies could not possibly transmit elemental qualities. Bernardus' confining of the quintessence to the firmament may perhaps represent a compromise between these positions. He was no doubt also influenced by Calcidius, *Commentarius* 129, pp. 171–72, glossing *Timaeus* 40D, who distinguishes the "serene" fire of the highest levels of the universe from the grosser ethereal fire; by Macrobius' account of the fiery nature of soul, *In Somn. Scip.* 1. 14. 17–18, p. 58; and perhaps by Cicero's presentation of the Stoic view of heavenly fire, *De natura deorum* 2. 21. 55–60, 2. 24. 64.

9. The source of this conception is *Asclepius* 19, p. 54, where παντόμορφος or *Omniformis*, "who fashions diverse forms for the diverse species," is said to be chief of the planetary usiarchs. On Bernardus' equation of the usiarch of the *Asclepius* with "genius" see Robert B. Woolsey, "Bernard Silvester and the Hermetic Asclepius," *Traditio*, 6 (1948), 343.

The glossator of Laud Misc. 515 observes of this figure that "he is called Oyarses or 'deputed power' because power over natural things has been deputed to him by the supreme God. He is also called 'genius' from 'generation,' because it is in accordance with the movement of this sphere (i.e. the firmament) that the natural generation of all things comes to pass. It may be seen then that this god is nothing else than the natural carrying out of those things which come through the activity of the heavens." This gloss clearly reinforces the suggestion of astral determinism in Bernardus' cosmology, and this genius figure is cited by Lemay, *Abu Ma'shar*, pp. 269–71, as evidence of the non-Platonic character of Bernardus' theory of forms. There is, however, ample evidence of the Platonic view elsewhere (e.g., in the work of Noys in 1.2, BW, p. 11. 92–94, and it is further possible that Bernardus' use of the genius figure here is intended to dramatize the Chartrian doctrine of secondary forms, intermediary images which link cosmic life with the divine ideas. See above, Book One, n. 35; Introduction, n. 166.

10. This and the following sentence are taken from *Asclepius* 35, p. 75, where the relation of sensible form and *species* to their divine source is discussed.

11. This is the attitude in which Geometry is first presented in Martianus, *De nuptiis* 6. 586, p. 291.

12. This sentence and the poem which follows are based on Apollo's reception of Mercury and Virtue in *De nuptiis* 1. 20–22, pp. 15–16.

13. *Quadret opus.* On the theme of the perfect *homo quadratus* in the twelfth century, see Edgar de Bruyne, *Études d'esthétique médiévale* (Bruges, 1946), II, 357–65.

14. Urania's role, as defined both by her self-characterization in this chapter and by her role in the creation of man, seems, like that of the celestial genius of the previous chapter, to combine elements of astral determinism and traditional Platonism. She seems clearly to include astrology among her resources, but Lemay goes too far in claiming that she is a wholly astrological conception, *Abu Ma'shar*, pp. 281–83. See above, Introduction, Section 6.

15. On the descent of the soul into bodily life see Macrobius, *In Somn. Scip.* 1. 14. 4–14, pp. 56–57.

16. *ut sit prudentior.* On the relation of cosmic knowledge to *prudentia* in earthly affairs see Calcidius, *Commentarius* 180, pp. 208–9; Macrobius, *In Somn. Scip.* 2. 17, pp. 151–54.

17. Cp. Boethius, *De consolatione philosophiae* 3, metr. 11; 5, metr. 4.

18. This phrase echoes Martianus' description of the abode of Pallas, *De nuptiis* 1. 39, p. 24.

19. On "Tugaton" (ταγαζόν, Plato's "Good") see Macrobius, *In Somn. Scip.* 1. 2. 14, p. 6; Silk, p. 158.

20. This description recalls that of the head of Arithmetic in Martianus, *De nuptiis* 7. 728, p. 365.

21. This description of the Trinity recalls the account of the procession of the Word in pseudo-Dionysius, *Celestial Hierarchy* 12 (trans. Eriugena, *PL*, 122, 246); the final phrase echoes Trismegistus' account, *Asclepius* 19, p. 53, of the difficulty of descriptions of the divine which, if not followed with the utmost attention, elude one's grasp, "fly or flow away or, better, flow back and reabsorb themselves in the fountain of light which is their source."

22. The language recalls Philology's prayer in the Empyrean, *De nuptiis* 2. 201–5, pp. 76–77.

23. Cp. Cicero, *De natura deorum* 2. 25. 64.

24. This account of Jove's dwelling is based on Martianus, *De nuptiis* 1. 64–68, pp. 30–32.

25. See Calcidius' account of the Fates, *Commentarius* 144, pp. 182–83, where Clotho is assigned dominion in the planetary spheres and said to be the power "through whom come about those things which the errant ('devius') motion of varying nature brings to pass."

26. See Martianus, *De nuptiis* 2. 166, 195, pp. 69, 75, on this river which issues from Mars and, with the other streams of planetary influence, descends to infuse the grove of Apollo.

27. The grouping of brilliance, power, and majesty suggests the "trinity" of the sun's substance, splendor, and heat, traditionally compared to the Father, Son, and Spirit; see, e.g., Honorius, *Elucidarium,* 1. 1, *PL,* 172, 1110–11. The same three attributes are enumerated by Remigius, comparing the sun to the world soul in his commentary on Boethius, *De consolatione,* ed. Silk, *op. cit.,* p. 335.

28. Cp. Macrobius, *In Somn. Scip.* 1. 20. 1–7, pp. 78–79.

29. With this description cp. Philology's prayer to the sun in *De nuptiis* 2. 185–93, pp. 73–74; Eriugena, *Annotationes in Marcianum,* p. 10, on the relation of the sun to the world soul; Alain, *Sermo de sphera intelligibili,* p. 299, on the "palace" of the world soul; *Asclepius* 29, pp. 67–68. Lemay, *Abu Ma'shar,* pp. 191–93, discusses the twelfth-century tendency to identify the rational "souls" of the planets with the nucleic influence of the sun.

30. Apollo's progeny are glossed as follows in Ms. Laud Misc. 515: " 'Harmless Phaethon,' or temperate heat, and 'Fruit of the Spring,' or annual fructification, are both born of the proper activity of the sun. Psyche, the animating principle of growth (*vegetatio*) in created life, takes rise from the sun. Swiftness, the swift lifetime of things, is produced by the revolution of the sun and the planets." "Celeritas" ("solis filia") and "Veris fructus" appear in Martianus' catalogue of gods, *De nuptiis* 1. 50, 53, p. 28.

31. On Psyche's work see above, Introduction, Section 6, and Cicero, *De natura deorum* 2. 21. 57, on that "ignis artificiosus" which is "teacher of the other arts."

32. On the vexed question of the relation of the orbits of Venus, Mercury, and the sun as understood by twelfth-century astronomy, see Gregory, *Anima mundi,* pp. 223–24.

33. See on this subject E. R. Curtius, *European Literature and the Latin Middle Ages,* trans. Willard Trask (New York, 1953), pp. 113–17.

34. See Lemay, *Abu Ma'shar,* pp. 253–54; Abu Ma'shar, *Introductorium,* 3. 4, f. clr; Hermann, *De essentiis,* p. 52.

35. Cp. Hugh of St. Victor, *Didascalicon,* 1. 7, pp. 14–15.

36. With this account of the nature and function of the moon cp. Macrobius, *In Somn. Scip.* 1. 19, 12–13, p. 75; 1. 21. 33–35, pp. 90–91.

37. Portions of this and the following sentence echo pseudo-Aristotle, *De mundo* 2, pp. 138–39.

38. This term, which I have found in no other work, seems to be a coinage based on some such scheme as that of *Asclepius* 7, p. 42, where

man is said to be a compound of οὐσιώδης, a form modeled on the divine, and a fourfold material principle, ὑλιχὸν; ;or that of Macrobius, *In Somn. Scip.* 1. 14. 7, p. 56, where λογιχόν, αισθητιχόν, and φυτιχόν are listed as the principles of reason, perception, and growth with which the soul is endowed. Cp. Bernardus' impressionistic use of "elementans" and "elementata" in 1. 4.

39. Silverstein has traced this allusion, which does not seem to correspond to any precise passage in Ptolemy's writings, to Abu Ma'shar; "Fabulous Cosmogony," p. 96, n. 27. Abu Ma'shar's influence seems to have made Ptolemy a name to conjure with; *Introductorium* 1. 7, f. a8v, refers to the Almagest, "in which the whole truth of universal wisdom ('universalis sapientie veritas integra') is contained." The glossator of Laud 515 cites Ptolemy in connection with the gloss on Bernardus' celestial Genius quoted above, Book Two, n. 9, where the influence of an astral determinism like that expounded by Abu Ma'shar seems clear.

40. This description is based in part on Martianus, *De nuptiis* 2. 170, p. 70.

41. See Macrobius, *In Somn. Scip.* 1. 14. 15, p. 58.

42. Cp. the host of the followers of Mercury which greets Philology as she rises through the spheres, *De nuptiis* 2. 170, p. 71.

43. An echo of *Asclepius* 4, p. 39.

44. For a similar conception see Ch. 4 of the pseudo-Aristotelian *Liber de causis*, ed. Otto Bardenhewer, *Ueber das reine Gute* (Freiburg im Breisgau, 1882), pp. 166–68. On the likelihood that this work was accessible in Latin in Bernardus' day see Silverstein, "Fabulous Cosmogony," p. 97, n. 31.

45. I.e., Seraphim.

46. This phrase and other details of the paragraph recall Calcidius, *Commentarius* 176, pp. 204–5, which is an account of the role of Noys or Providence.

47. Cp. Guillaume de Conches, *De philosophia mundi* 1. 16, *PL,* 172, 47.

48. See *Philippians* 4. 7.

49. Cp. Guillaume de Conches, *De philosophia mundi* 1. 17, *PL,* 172, 47.

50. Cp. Apuleius, *De deo Socratis* 6, pp. 13–14.

51. Cp. Martianus *De nuptiis* 2. 152, p. 65.

52. Cp. Apuleius, *De deo Socratis* 6, p. 14; Honorius, *Elucidarium* 2. 28, *PL,* 172, 1154.

53. Cp. Calcidius, *Commentarius* 132, 195, pp. 174–75, 217.

54. Cp. Apuleius, *De deo Socratis* 6, 17, pp. 14, 26–27; Censorinus, *De die natali* 3, ed. O. Jahn (Berlin, 1845), p. 7; Martianus, *De nuptiis* 2. 151, p. 65.

55. Cp. Martianus, *De nuptiis* 2. 154–55, pp. 65–66.

56. With this sentence and the rest of the paragraph cp. Martianus, *De nuptiis* 2. 167, p. 69.

57. See Martianus, *De nuptiis* 2. 161, p. 68.

58. This section of the *Microcosmus* seems to be broadly modeled on Firmicus Maternus' vindication of astrology, *Mathesis* 1. 4, pp. 11–14, as well as on the discourse of Pythagoras in Ovid, *Metamorphoses* 15. 75–478 (see above, Introduction, Section 7).

59. For a similar use of *alteritas* see Thierry, *De sex dierum operibus*, p. 196, quoted by Silverstein, "Fabulous Cosmogony," p. 101, on the relation of created existence to God.

60. With this sentence cp. Calcidius, *Commentarius* 199, pp. 219–20, and *Timaeus* 42CD.

6.. With this and the previous sentence cp. Silk, p. 178: "Through matter existence is altered, through form its name endures. On the basis of this notion the name of world is assigned to the form of the world which is beheld, and yet this possesses only the name of world and is not the true world. . . . Its form is imposed upon fluctuating matter and that which we now behold is given expression and assigned an abiding name. Existence is only altered, for those things which perish for want of sustenance in this (i.e. the immediate, visible) world are made to live on in that world (i.e. on the level of ceaseless flux) through the power of generation." With the final sentence cp. the passage from *Cosmographia*, 2. 14, quoted above, Introduction, pp. 51–52.

Alain de Lille plays on these lines in *Anticlaudianus* 1. 365–71, ed. Robert Bossuat (Paris, 1955), pp. 67–68, where Reason stresses the superiority of spirit to flesh: "the one knows earthly, the other heavenly things; this has its dwelling on high, that in the world. One is forced to pay tribute to death, but the law of death exempts the other. One abides, the other flows away, one endures, the other perishes; one enjoys but the name of being, the other the power . . ." For Alain and the school of "Porretan" theologians to which he belongs, the impress of form is the *osculum* of the divine wisdom, a direct manifestation of God, while the conception presented by Bernardus, though finally, I think, capable of being understood in emanationist terms (see above, Introduction, Section 7), seems also to reflect in its terminology the Arabic version of Aristotelian physics. See Lemay, *Abu Ma'shar*, pp. 175–76.

62. Cp. Macrobius, *In Somn. Scip.* 2.12.13–14, pp. 132–33, which probably provided the basis for the passage quoted from Silk in the preceding note. It is noteworthy for its citation of Plotinus as authority for this view of matter.

63. I.e., the atmosphere. See Vergil, *Aeneid* 1. 50–63.

64. Discussing the five "regions" of the universe Calcidius, *Commentarius* 129, pp. 171–72, assigns the name of "hygran usian" to the realm of the

lower atmosphere immediately adjacent to the earth. Bernardus' intention in appropriating the term is perhaps to impute to Paradise a purer, less gross materiality than that of earthly nature. Similarly Eriugena sees as a consequence of the Fall the coarsening of man's bodily nature, which originally had been more refined; *De divisione naturae* 2. 25, *PL*, 122, 582.

65. Cp. Hermann, *De essentiis*, p. 62, who compares the work of medicine to that of nature, in that it "compounds the substance of the human body out of natural things."

66. Hermann, *ibid.*, p. 63, notes that medicine also employs natural substances in ways which are contrary to nature.

67. This ray of light is the first of several hints that Physis, like Nature, is guided by "sacred and blessed instincts." Thus her shaping of the human head is said to conform to an inscrutable design (2. 12, BW, p. 64), and its appropriateness as the seat of wisdom is characterized as more or less an article of faith (2. 14, BW, p. 65).

68. Cp. Silk, pp. 155–56.

69. Cp. *Asclepius* 8, pp. 43–44. See also Woolsey, "Bernard Silvester and the Hermetic Asclepius," pp. 343–44.

70. On the rich tradition associated with the image of man presented in this sentence see Silverstein, "Fabulous Cosmogony," p. 97, n. 28; Southern, *Medieval Humanism*, pp. 39–40. Like 2. 8, the discourse which follows seems to recall Firmicus, *Mathesis* 1. 4, pp. 11–14.

71. With this and the following sentences, cp. *Asclepius* 8, pp. 43–44.

72. Death is described in these terms in *Asclepius* 27, pp. 65–66.

73. With this sentence cp. Firmicus, *Mathesis*, 1. 4. 3, p. 11. The use of "guest" ("hospes") recalls the ironic reference to the brief hospitality enjoyed by man in Eden (1. 3, BW, p. 25. 333–34) and so helps to confirm Bernardus' association of the platonic myth of the soul's return with the recovery of man's paradisal state.

74. Cp. *Timaeus* 41CD, 42E. On the "engendering" of the soul ("semens animae") cp. Calcidius, *Commentarius* 141, p. 181.

75. The most likely source of this motif is the "speculum Uraniae" (in most manuscripts "Aniae") bestowed upon Psyche by Sophia in *De nuptiis* 1. 7, p. 8, which confers upon the soul the capacity for self-knowledge and the consequent desire to recover its original state. That Bernardus assigns such a "speculum" to his Urania suggests that he knew a manuscript of the *De nuptiis* which preserved this reading, though in the commentary on Martianus he discusses only the "Aniae" reading.

The three gifts, the Mirror, the Table and the Book, may perhaps be taken as an ahistorical counterpart to the idea of the stages of revelation, natural, scriptural, and Christian as presented by Hugh of St. Victor in the *De Sacramentis*. See above, Introduction, Section 3. In Laud Misc. 515 they are glossed as theology, astronomy, and physics.

76. This is the most explicit of the various hints which associate Ende-

lechia with the "ignis artifex." See Gregory, *Platonismo medievale*, p. 137.

77. This account of the roles of the Fates recalls Calcidius, *Commentarius* 144, pp. 182–83.

78. See above, 2. 5.

79. Cp. Firmicus, *Mathesis* 3. Proem. 2–3, pp. 90–91; *De sex rerum principiis* 447, p. 289.

80. On these principles see *Timaeus* 36B–37B; Calcidius, *Commentarius* 53, p. 101; Thierry, *De sex dierum operibus*, p. 196. Silverstein, "Fabulous Cosmogony," pp. 101–2.

81. Cp. Hermann, *De essentiis*, pp. 39–40; Gundisalvus, *De anima* 4, 7, ed. Muckle, pp. 45–46, 53. See also Eriugena, *De divisione naturae* 1. 34, *PL*, 122, 479.

82. Cp. *Timaeus* 49–51; Calcidius, *Commentarius* 308–9, 317, pp. 309–10, 313.

83. See *Timaeus* 51B, 53B; Calcidius, *Commentarius* 337, 354, pp. 330–31, 344–45.

84. Cp. Hermann, *De essentiis*, pp. 43–44.

85. On "complexion" see *Timaeus* 31B–32C; Macrobius, *In Somn. Scip.* 1. 6. 24–33, pp. 22–24; Constantinus Africanus, *De communibus medico cognitu necessariis locis* 1. 3–5, in *Opera* (Basel, 1539), II, 5–8; Guillaume de Conches, *Glosae super Platonem* 59, pp. 129–30.

86. Cp. Constantinus, *De communibus locis* 1. 5, p. 8.

87. See *Timaeus* 44DE; Calcidius, *Commentarius* 232, p. 246.

88. Cp. Calcidius, *Commentarius* 231, p. 245.

89. Cp. *ibid.* 233, p. 247.

90. Cp. *ibid.* 231, p. 245.

91. On the extensive use of this threefold conception of human psychology see Silverstein, "Fabulous Cosmogony," pp. 97–98 and n. 34; also my *Platonism and Poetry*, pp. 94–97, 116–18.

92. Cp. Calcidius, *Commentarius* 231, p. 245.

93. The phrase "messenger senses" ("internuntia sentiendi") is apparently borrowed from Apuleius, *De Platone et eius dogmate* 1. 16, p. 100.

94. See Calcidius, *Commentarius* 246, p. 257; Constantinus, *De communibus locis* 4. 11, p. 92; Guillaume, *De philosophia mundi* 4. 26, *PL*, 172, 96.

95. With this and the following two sentences cp. Calcidius, *Commentarius* 244, 247–48, pp. 255, 258–59.

96. Cp. *Timaeus* 45DE.

97. Cp. Constantinus, *De communibus locis* 3. 12, p. 59.

98. Cp. *ibid., loc. cit.;* Calcidius, *Commentarius* 246, p. 257, speaks of a fourfold covering.

99. Cp. Guillaume, *De philosophia mundi* 4. 25, *PL*, 172, 91–92.

100. Cp. Calcidius, *Commentarius* 266, p. 271: "It is contemplation which guides men's minds to the very vault of the firmament, wherefore

Anaxagoras, when he was asked why he had been born, is said to have replied, gesturing toward the sky and pointing out the stars, 'to contemplate all these.' " See also pseudo-Guillaume de Conches, *Philosophia seu summa philosophiae,* ed. Carmelo Ottaviano, *Un brano inedito della "Philosophia" di Guglielmo di Conches* (Naples, 1935), p. 22.

101. Cp. Constantinus, *De communibus locis* 3. 15, p. 62.

102. Cp. Cicero, *De natura deorum* 2. 58. 146.

103. Panchaia was a fabulous island, famous for incense and myrrh. See Vergil, *Georgics* 2. 139, 4. 379.

104. Cp. Calcidius, *Commentarius* 232, p. 246.

105. Cp. *ibid.* 224, p. 239.

106. Cp. *ibid.* 187, p. 212.

107. Cp. Constantinus, *De communibus locis* 5. 35, p. 139.

108. Silverstein, "Fabulous Cosmogony," p. 109, n. 118, takes for granted that the genii mentioned here are the tutelary spirits of marriage, and cites the parallel of Censorinus, *De die natali* 3, ed. Jahn, p. 7. The association does not seem to me by any means so clear, and I think the genii are to be associated with man's procreative instincts and "life-giving weapons" ("genialibus armis"), rather than with any formal consecration of these. Stock, *Myth and Science,* pp. 218–19, suggests that they stand for "the masculine and feminine aspects of creativity latent in matter."

It is further possible that in introducing the genius figure at this point Bernardus is exploiting the "impious" notion imputed by Calcidius to certain Stoic thinkers, which identifies the divine with Silva, or with a quality of Silva, and conceives it as moving through Silva "as semen through the genitalia," so that it becomes the immediate as well as the ultimate cause of life; see *Commentarius* 294, pp. 296–97. Such a conception might be taken as a metaphor for Eriugena's all-pervading "motus" toward fulfillment.

109. With this and the following two sentences cp. the fifth Elegy of Maximianus, lines 110 ff., on the creative power of the phallus.

110. Cp. Constantinus, *De communibus locis* 3. 35, pp. 77.

111. Cp. *ibid.* 3. 34, p. 77.

112. The language of this and the following sentence is close to that of Calcidius, *Commentarius* 24, p. 75; cp. *Commentarius* 192, p. 215, and *Timaeus* 32C.

113. Cp. *Timaeus* 33B–34A.

# Bibliography

## Primary Sources

Abelard, Peter. *Opera. Patrologia Latina (PL)*, Vol. 178.

Abu Ma'shar. *Maius introductorium in astronomiam*. Venice, 1506.

Adelhard of Bath. *De eodem et diverso*, ed. Hans Willner. *Beiträge zur Geschichte der Philosophie des Mittelalters*, 4, No. 1 (1903).

Alain de Lille. *Anticlaudianus*, ed. Robert Bossuat. Paris, 1955.

——. *De planctu naturae*, ed. Thomas Wright. In *Anglo-Latin Satrical Poets of the Twelfth Century* (2 vols., London, 1872), II, 429–522.

——. *Textes inédits*, ed. Marie-Thérèse d'Alverny. Paris, 1965.

Andreas Capellanus. *De amore*, ed. Ernst Trojel. Copenhagen, 1892.

Apuleius. *De philosophia libri*, ed. Paul Thomas. Leipzig, 1908.

Bernard of Clairvaux. *Sermones in Cantica Canticorum. PL*, 183, 785–1198.

Bernardus Silvestris. Commentary on Martianus Capella. MS, Cambridge, University Library, Mm. 1.18, ff. 1r–28r.

——. *Commentum super sex libros Eneidos Virgilii*, ed. Wilhelm Riedel. Greifswald, 1924.

——. *Cosmographia*, ed. André Vernet. *Bernardus Silvestris: Recherches sur l'auteur et l'oeuvre suivies d'une édition critique de la "Cosmographia."* Diss., Paris, 1938.

——. *Cosmographia*. MS, Oxford, Bodleian, Laud Misc. 515, ff. 182r–219r.

——. *De mundi universitate (Cosmographia)*, ed. C.S. Barach, J. Wrobel. Innsbruck, 1876.

——. *Experimentarius*, ed. Mirella Brini-Savorelli. *Rivista critica di storia della filosofia*, 14 (1959), 283–342.

——. *Mathematicus*, ed. Barthélemy Hauréau. Paris, 1895.

——. *Mathematicus. PL*, 171, 1365–80.

Boethius. *De institutione arithmetica*, ed. G. Friedlein. Leipzig, 1867.
——. *De Trinitate*, ed. H.F. Stewart, E.K. Rand. In *Boethius: The Theological Tractates and the Consolation of Philosophy* (New York, 1918), pp. 2–31.
Calcidius. *Commentarius in Timaeum Platonis*, ed. J.H. Waszink. London, 1962.
Censorinus. *De die natali liber*, ed. Otto Jahn. Berlin, 1845.
Constantinus Africanus. *Opera.* 2 vols. Basel, 1539.
Dominicus Gundisalvus (Gundissalinus). *De processione mundi*, ed. Georg Bülow. *Beiträge zur Geschichte der Philosophie des Mittelalters*, 24, No. 3 (1925).
——. *De anima*, ed. J. T. Muckle, *Medieval Studies*, 2 (1939), 23–103.
Eberhard. *Laborintus*, ed. Edmond Faral. *Les arts poétiques du xiie et du xiiie siécle* (Paris, 1924), pp. 336–77.
Firmicus Maternus. *Mathesis*, ed. W. Kroll, F. Skutsch. 2 vols. Leipzig, 1897.
Fulgentius. *Opera*, ed. Rudolph Helm. Leipzig, 1898.
Geoffroi de Vinsauf. *Poetria nova*, ed. Edmond Faral. *Les arts poétiques du xiie et du xiiie siécle* (Paris, 1924), pp. 194–262.
Gervais of Melkley. *Ars poetica*, ed. Hans Jürgen Gräbener. Munster, 1965.
Guillaume de Conches. *De philosophia mundi. PL*, 172, 39–102.
——. *Glosae super Platonem*, ed. Edouard Jeauneau. Paris, 1965.
——. Glosses on Boethius, ed. Charles Jourdain. "Des commentaires inédits de Guillaume de Conches et de Nicolas Triveth sur la Consolation de la philosophie de Boèce." *Notices et extraits des manuscrits de la Bibliothèque Impériale*, 20.2 (1862), 40–82.
Hermann of Carinthia. *De essentiis*, ed. P.M. Alonso. *Miscellanea Comillas*, Vol. 5 (1946).
Honorius Augustodunensis. *Opera. PL*, Vol. 172.
Hugh of St. Victor. *Opera. PL*, Vols. 175–77.
——. *Didascalicon*, ed. Charles H. Buttimer. Washington, 1939. (Catholic University of America. Studies in Medieval and Renaissance Latin, Vol. 10.)
——. *The Didascalicon of Hugh of St. Victor*, trans. Jerome Taylor. New York, 1961.
Jean de Hanville. *Architrenius*, ed. Thomas Wright. In *Anglo-Latin Satirical Poets of the Twelfth Century* (2 vols., London, 1872), I, 240–391.
Johannes Scotus Eriugena. *Opera. PL*, Vol. 122.
——. *Annotationes in Marcianum*, ed. Cora E. Lutz. Cambridge, Mass., 1939.
John of Salisbury. *Metalogicon*, ed. C.C.J. Webb. Oxford, 1929.
Macer Floridus. *De viribus herbarum*, ed. Louis Choulant. Leipzig, 1832.
Marcrobius. *Opera*, ed. James Willis. 2 vols. Leipzig, 1863.

Mangegold of Lautenbach. *Opusculum contra Wolfelmum.* PL, 155, 149–176.

Martianus Capella. *De nuptiis Philologiae et Mercurii,* ed. Adolph Dick. Leipzig, 1925.

Peter the Chanter. *Verbum abbreviatum.* PL, 205, 21–370.

Peter Damian. *De divina omnipotentia.* PL, 145, 595–622.

Remigius of Auxerre. *Commentum in Martianum Capellam, Libri I–II,* ed. Cora E. Lutz. Leiden, 1962.

Thierry of Chartres. Commentary on Boethius, *De Trinitate,* ed. N.M. Häring. *Archives d'histoire doctrinale et littéraire du moyen âge,* 31 (1956), 257–325.

——. *De sex dierum operibus,* ed. N.M. Häring. In "The Creation and Creator of the World according to Thierry of Chartres and Clarenbaldus of Arras," *Archives d'histoire doctrinale et littéraire du moyen âge,* 30 (1955), 184–200.

——. *Prologus in Eptateuchon,* ed. Edouard Jeauneau. *Medieval Studies,* 16 (1954), 171–75.

*Asclepius.* In Apuleius. *De philosophia libri,* ed. Paul Thomas. Leipzig, 1908.

*De mundi constitutione.* PL, 90, 881–910.

*De septem septenis.* PL, 199, 945–964.

*Liber de causis,* ed. Otto Bardenhewer. *Ueber das reine Gute.* Freiburg im Breisgau, 1882.

*Metamorphosis Goliae episcopi,* ed. Thomas Wright. *Latin Poems commonly attributed to Walter Mapes* (London, 1841), pp. 21–30.

*Saeculi noni auctoris in Boetii consolationem philosophiae commentarius,* ed. E.T. Silk. Rome, 1935. (American Academy in Rome. Papers and Monographs, Vol. 9.)

## Secondary Sources

Adler, Alfred. "The *Roman de Thebes,* a 'Consolatio Philosophiae,'" *Romanische Forschungen,* 72 (1960), 257–76.

Allen, Judson B. *The Friar as Critic.* Nashville, 1971.

Allers, Rudolph. "Microcosmus from Anaximandros to Paracelsus," *Traditio,* 2 (1944), 319–407.

d'Alverny, Marie-Thérèse. "Le cosmos symbolique du xiie siècle," *Archives d'histoire doctrinale et littéraire du moyen âge,* 28 (1953), 31–81.

——. "Alain de Lille et la *Theologia.*" In *L'homme devant Dieu: Mélanges offerts au Père Henri de Lubac* (3 vols., Paris, 1964), II, 111–28. See also, above, Alain de Lille.

Baron, Roger. *Science et sagesse chez Hugues de St.-Victor.* Paris, 1957.

Bennett, J.A.W. *The Parlement of Foules*. Oxford, 1957.

Bezold, Friedrich von. *Das Fortleben der antiken Götter im mittelalter-lichen Humanismus*. Bonn, 1922.

Bezzola, Reto R. *Les origines et la formation de la littérature courtoise en occident*. 3 vols. in 4 Paris, 1944–63.

Bloch, Marc. *Feudal Society*, trans. L.A. Manyon. Chicago, 1961.

Bolgar, R.R. *The Classical Heritage and Its Beneficiaries*. Cambridge, 1954.

Brini-Savorelli, Mirella. See, above, Bernardus Silvestris.

Brinkmann, Hennig. "Wege der epischen Dichtung im Mittelalter," *Archiv für das Studium der neueren Sprachen*, 200 (1963–64), 401–35.

de Bruyne, Edgar. *Études d'esthétique médiévale*. 3 vols. Bruges, 1946. (Rijksuniversiteit te Gent. Werken Uitgegeven door de Faculteit van de Wijsbegeerte en Letteren. Afl. 97–99.)

Chenu, Marie-Dominique. *La théologie au douzième siècle*. Paris, 1957.

——. *Nature, Man and Society in the Twelfth Century*, trans. Jerome Taylor and L.K. Little. Chicago, 1968. (Selected chapters from *La théologie au douzième siècle*.)

Clerval, J. Alexandre. *Les écoles de Chartres au moyen âge*. Chartres, 1895.

Courcelle, Pierre. "Étude critique sur les commentaires de la Consolation de Boèce (ixe–xve siècles)," *Archives d'histoire doctrinale et littéraire du moyen âge*, 12 (1939), 5–140.

——. *La consolation de philosophie dans la tradition littéraire*. Paris, 1967.

Curtius, E. R. *European Literature and the Latin Middle Ages*, trans. Willard Trask. New York, 1953.

Delhaye, Philippe. " 'Grammatica' et 'ethica' au xiie siècle," *Recherches de théologie ancienne et médiévale*, 25 (1958), 59–110.

——. *Le Microcosmos de Godefroy de St.-Victor: Étude théologique*. Lille, 1951.

Dronke, Peter. "L'amor che move il sole e l'altre stelle," *Studi medievali*, 6 (1965), 389–422.

——. *Medieval Latin and the Rise of European love-lyric*. 2 vols. Oxford, 1965.

——. "New Approaches to the School of Chartres," *Anuario de estudios medievales*, 6 (1969), 117–40.

Economou, George. *The Goddess Natura in Medieval Literature*. Cambridge, Mass., 1972.

Faral, Edmond. *Les arts poétiques du xiie et du xiiie siècle*. Paris, 1924.

——. "Le manuscrit 511 du 'Hunterian Museum' de Glasgow," *Studi medievali*, 9 (1936), 18–119.

——. " 'Le Roman de la Rose' et la pensée française au xiiie siècle," *Revue des deux mondes*, 35 (1926), 430–57.

Frappier, Jean. "Vues sur les conceptions courtoises dans les littératures d'oc et d'oïl au xiie siècle," *Cahiers de civilization médiévale*, 2 (1959), 135–56.

Gandillac, Maurice de. "Place et signification de la technique dans le monde médiéval." In *Tecnica e casistica: Atti del convegno indetto dal Centro Internazionale di Studi Umanistici e dal'Istituto di Studi Filosofici, Roma, 7–12 Gennaio, 1964* (Rome, 1964), pp. 265–75.

Garin, Eugenio. *Studi sul platonismo medievale*. Florence, 1958.

Gilson, Étienne. "La cosmogonie de Bernardus Silvestris," *Archives d'histoire doctrinale et littéraire du moyen âge*. 3 (1928), 5–24.

Gregory, Tullio. *Anima mundi. La filosofia di Guglielmo di Conches et la Scuola di Chartres*. Florence, 1955.

———. "L'idea di natura nella filosofia medievale prima del' ingresso della fisica di Aristotele: il secolo xii." In *La filosofia della natura nel Medioevo: Atti del Terzo Congresso Internazionale di Filosofia Medievale, 1964* (Milan, 1966), pp. 29–65.

———. *Platonismo medievale: studi e ricerche*. Rome, 1958.

Györy, Jean. "Le cosmos, un songe," *Annales Universitatis Scientiarum Budapestensis: Sectio Philologica*, 4 (1963), 87–110.

Haskins, Charles H. *The Renaissance of the Twelfth Century*. Cambridge, Mass., 1927.

———. *Studies in the History of Medieval Science*. Cambridge, Mass., 1924.

Hauréau, Barthélemy. "Mémoire sur quelques chanceliers de l'église de Chartres," *Mémoires de l'Académie des Inscriptions et Belles-Lettres*, 31.2 (1884), 63–122.

Huizinga, Johan. "Über die Verknupfung des Poetischen mit dem Theologischen bei Alanus de Insulis," *Mededeelingen der Koninklijke Akademie van Wetenschappen. Amsterdam*, 74B, No. 6 (1932), 89–199. Reprinted in Huizinga's *Verzamelde Werken* (9 vols., Haarlem, 1948–53), IV, 3–84.

Javelet, Robert. *Image et ressemblance au douzième siècle de St. Anselme à Alain de Lille*. 2 vols. Paris, 1967.

———. "Image de Dieu et nature au xiie siècle." In *La filosofia della natura nel Medioevo* (see Gregory, Tullio), pp. 286–96.

Jeauneau, Edouard. "Macrobe, source du platonisme chartrain," *Studi medievali*, 1 (1960), 3–24.

———. "Notes sur l'École de Chartres," *Studi medievali*, 5 (1964), 821–65.

———. "L'usage de la notion d'integumentum à travers les gloses de Guillaume de Conches." *Archives d'histoire doctrinale et littéraire du moyen âge*, 32 (1957), 35–100. See also, above, Guillaume de Conches.

Kelly, Douglas. "The Scope and Treatment of Composition in the Twelfth and Thirteenth-Century Arts of Poetry." *Speculum*, 41 (1966), 261–78.

Kranz, Walter. *Kosmos. Archiv für Begriffsgeschichte*, 2 (1955).

Krayer, Rudolf. *Frauenlob und die Natur-Allegorese: Motivgeschichtliche Unterschungen*. Heidelberg, 1960.

Kristeller, P. O. "Beitrag der Schule von Salerno zur Entwicklung der Scholastischen Wissenschaft im 12. Jahrhundert." In *Artes liberales: von der antiken Bildung zur Wissenschaft des Mittelalters*, ed. Josef Koch (Leiden and Cologne, 1959), pp. 84–90.

——. "The School of Salerno: Its Development and Its Contribution to the History of Learning," *Bulletin of the History of Medicine*, 17 (1945), 138–94. Reprinted in Kristeller's *Studies in Renaissance Thought and Letters* (Rome, 1956), pp. 495–551.

Le Goff, Jacques. *La civilization de l'Occident médiéval*. Paris, 1967.

——. *Les intéllectuels au moyen âge*. Paris, 1957.

——. "Métier et profession d'aprés les manuels de confesseurs au moyen-âge." In *Beiträge zum Berufsbewusstsein des mittelalterlichen Menschen*, ed. Paul Wilpert (Berlin, 1964), pp. 44–60. (*Miscellanea medievalia*, Vol. 3.)

Lemay, Richard. *Abu Ma'shar and Latin Aristotelianism in the Twelfth Century*. Beirut, 1962.

Liebeschütz, Hans. "Kosmologische Motive in der Bildungswelt der Frühscholastik." *Vorträge der Bibliothek Warburg, 1923–24* (Berlin, 1926), pp. 87–144.

——. *Medieval Humanism in the Life and Writings of John of Salisbury*. London, 1951.

de Lubac, Henri. *Exégèse médiévale*. 2 vols. in 4. Paris, 1959–64.

Luscombe, D. E. *The School of Peter Abelard*. Cambridge, 1970.

McKeon, Richard. "Medicine and Philosophy in the Eleventh and Twelfth Centuries: The Problem of Elements," *The Thomist*, 24 (1961), 211–56.

——. "Poetry and Philosophy in the Twelfth Century: The Renaissance of Rhetoric," *Modern Philology*, 43 (1945–46), 217–34. Reprinted in *Critics and Criticism, Ancient and Modern*, ed. R. S. Crane (Chicago, 1952), pp. 297–318.

Munari, Franco. "Medievalia," *Philologus*, 105 (1960), 279–92.

Nuchelmans, Gabriel. "Philologie et son mariage avec Mercure jusqu'à la fin du xiie siècle," *Latomus*, 16 (1957), 84–107.

Padoan, Giorgio. "Tradizione e fortuna del commento all'*Eneide* di Bernardo Silvestre," *Italia medioevale e umanistica*, 3 (1960), 227–40.

Parent, J. M. *La doctrine de la création dans l'école de Chartres*. Paris and Ottawa, 1938.

Pedersen, Olaf. "Du Quadrivium à la Physique." In *Artes liberales* (see above, P. O. Kristeller), pp. 107–23.

Pollman, Leo. *Chrétien von Troyes und der Conte del Graal*. Tübingen, 1965. (*Beihefte zur Zeitschrift für romanische Philologie*. Hft. 110.)

——. *Das Epos in den romanischen Literaturen*. Stuttgart, 1966.

Poole, R. L. "The Masters of the Schools at Paris and Chartres in John of Salisbury's Time." *English Historical Review*, 35 (1920), 321–42. Reprinted in Poole's *Studies in Chronology and History*, ed. A. L. Poole (Oxford, 1934), pp. 223–47.

Schipperges, Heinrich. "Einflüsse arabischer Medezin auf die Mikrokosmosliteratur des 12. Jahrhunderts." In *Antike und Orient im Mittelalter: Vorträge der Kölner Mediaevistenangung 1956–59*, ed. Paul Wilpert (Berlin, 1962), pp. 129–53. (*Miscellanea Medievalia*, Vol. 1.)

Silverstein, Theodore. "Elementatum: Its Appearance among the Twelfth-Century Cosmogonists." *Medieval Studies*, 16 (1954), 156–62.

———. "The Fabulous Cosmogony of Bernardus Silvestris." *Modern Philology*, 46 (1948–49), 92–116.

———. "Guillaume de Conches and Nemesius of Emessa: On the Sources of the 'New Science' of the Twelfth Century." In *Harry Austryn Wolfson Jubilee Volume* (3 vols., Jerusalem, 1965), II, 719–34.

Southern, R. W. *The Making of the Middle Ages*. New Haven, 1953.

———. *Medieval Humanism and Other Studies*. Oxford, 1970.

Stock, Brian. *Myth and Science in the Twelfth Century: A Study of Bernard Silvester*. Princeton, 1972.

Taylor, Jerome. See, above, Hugh of St. Victor.

Thorndike, Lynn. *A History of Magic and Experimental Science*. 8 vols. New York, 1923–58.

Tuzet, Hélène. *Le cosmos et l'imagination*. Paris, 1965.

Varvaro, Alberto. "Scuola e cultura in Francia nel xii secolo." *Studi mediolatini e volgari*, 10 (1962), 299–330.

Vernet, André. "Une épitaphe inédite de Thierry de Chartres." In *Recueil de Travaux offert à M. Clovis Brunel* (2 vols., Paris, 1955), II, 660–70.

Von den Steinen, Wolfram. *Der Kosmos des Mittelalters*. 2d ed. Bern, 1967.

———. *Menschen im Mittelalter*. Bern, 1967.

———. "Les sujets d'inspiration chez les poètes latins du xiie siècle," *Cahiers de civilization médiévale*, 9 (1966), 165–75, 363–83.

Waddell, Helen. *The Wandering Scholars*. 7th ed. London, 1934.

Wetherbee, Winthrop. "The Function of Poetry in the *De planctu naturae* of Alain de Lille," *Traditio*, 25 (1969), 87–125.

———. *Platonism and Poetry in the Twelfth Century*. Princeton, 1972.

Woolsey, Robert B. "Bernard Silvester and the Hermetic Asclepius," *Traditio*, 6 (1948), 340–44.

# Index

*(The index is in two parts. Asterisked items are listed in both sections.)*

## I. Themes, Images, and Technical Terms in the Cosmographia

## II. Index to Introduction and Notes